ALGEBRA

PATTERNS AND GRAPHS

Walker Maths Essentials: Algebra Patterns and Graphs 4+
1st Edition
Charlotte Walker
Victoria Walker

Cover designer: Cheryl Smith, Macarn Design
Text design: Cheryl Smith, Macarn Design
Production controller: Siew Han Ong

Any URLs contained in this publication were checked for currency during the production process. Note, however, that the publisher cannot vouch for the ongoing currency of URLs.

Acknowledgements
Cover photo courtesy of Shutterstock.

We wish to thank the Boards of Trustees of Darfield and Riccarton High Schools for allowing us to use materials and ideas developed while teaching. Our thanks also go to all past and present colleagues, especially Kath Wilson, who have generously shared their experience and ideas.

For product information and technology assistance,
in Australia call **1300 790 853**;
in New Zealand call **0800 449 725**

For permission to use material from this text or product, please email
aust.permissions@cengage.com

National Library of New Zealand Cataloguing-in-Publication Data
A catalogue record for this book is available from the National Library of New Zealand.

978 0 17 045150 5

Cengage Learning Australia
Level 7, 80 Dorcas Street
South Melbourne, Victoria Australia 3205

Cengage Learning New Zealand
Unit 4B Rosedale Office Park
331 Rosedale Road, Albany, North Shore 0632, NZ

For learning solutions, visit **cengage.co.nz**

Printed in China by 1010 Printing International Limited.
2 3 4 5 6 7 25

CONTENTS

Glossary

Make your own glossary of key terms:

Term	Definition	Picture/Example
Expression		
Like terms		
Variable		
Term		
Sequence		
Origin		
Coordinate		
Horizontal		
Vertical		
Discrete data		
Continuous data		
Linear		
Non-linear		

ISBN: 9780170451505

Substitution

- When substituting, you replace **variables** (letters) with **numbers**.
- Sometimes you will need to apply BEDMAS.

Examples: If $n = 4$, find the values of A.

1 $A = n + 5$
$= \mathbf{4} + 5$
$A = 9$

We replace the n with **4**.

2 $A = 2n$
$= 2 \times \mathbf{4}$
$A = 8$

Remember, if we can't see an operator (sign), there is a **x**.

We replace the n with **4**.

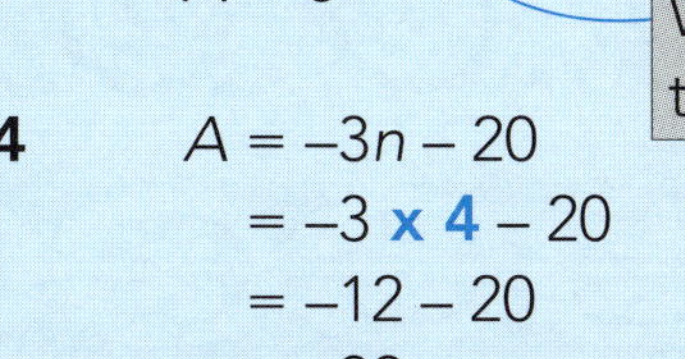

3 $A = 5n - 10$
$= 5 \times \mathbf{4} - 10$
$= 20 - 10$
$= 10$

4 $A = -3n - 20$
$= -3 \times \mathbf{4} - 20$
$= -12 - 20$
$= -32$

Complete the following table.

Formula	$n = 3$	$n = 7$	$n = 10$
$A = n + 3$	$A =$ ______ $=$ ______	$A =$ ______ $=$ ______	$A =$ ______ $=$ ______
$A = 5n$	$A =$ ______ $=$ ______	$A =$ ______ $=$ ______	$A =$ ______ $=$ ______
$A = 6n - 4$	$A =$ ______ $=$ ______ $=$ ______	$A =$ ______ $=$ ______ $=$ ______	$A =$ ______ $=$ ______ $=$ ______
$A = -2n + 1$	$A =$ ______ $=$ ______ $=$ ______	$A =$ ______ $=$ ______ $=$ ______	$A =$ ______ $=$ ______ $=$ ______
$A = -3n - 1$	$A =$ ______ $-$ ______ $=$ ______	$A =$ ______ $-$ ______ $=$ ______	$A =$ ______ $-$ ______ $=$ ______

Patterns

- Patterns can be seen all around us, often in art and cultural decorations.

 ISBN: 9780170451505

Continuing patterns

- You need to work out how patterns change so you can draw more shapes.

Draw the next shape for each of these patterns.

1

2

3

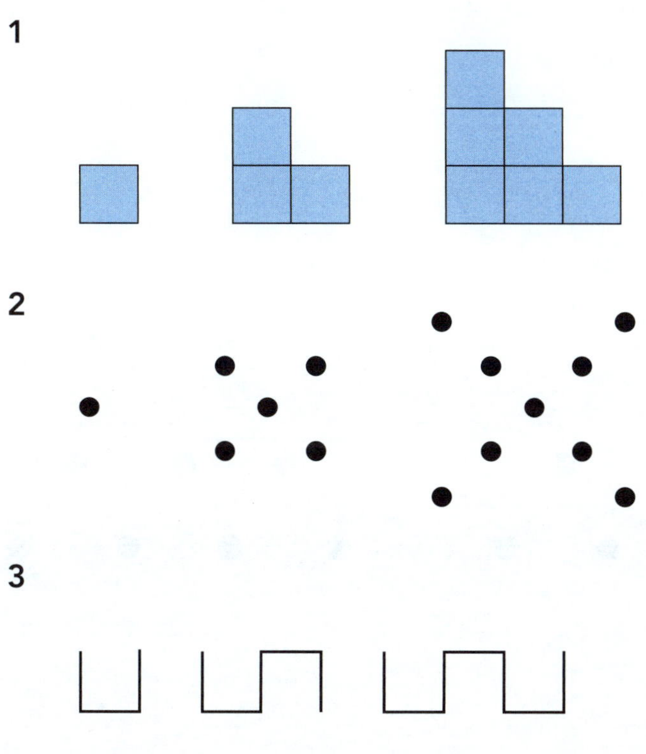

4

5

6

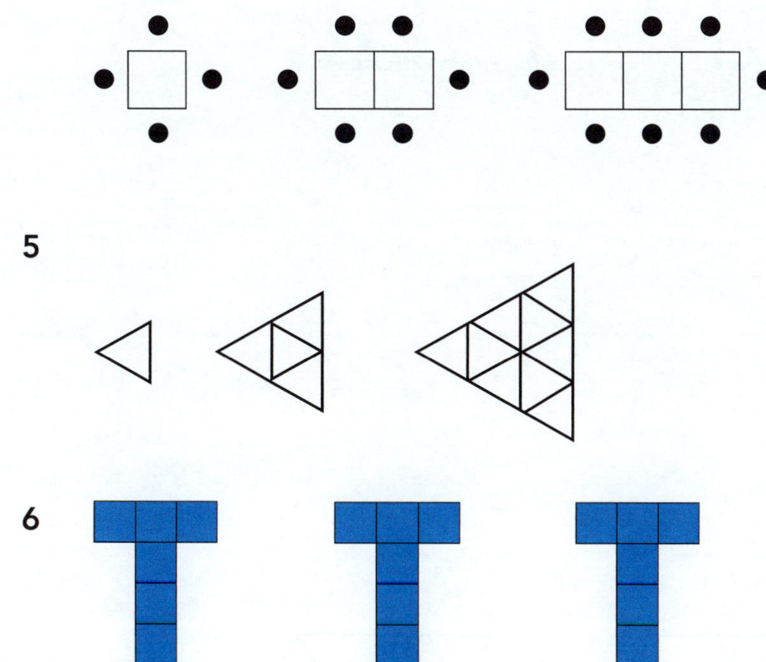

ISBN: 9780170451505

From pictures to numbers

Write down the number of objects in each of these shapes.

1 Number of squares

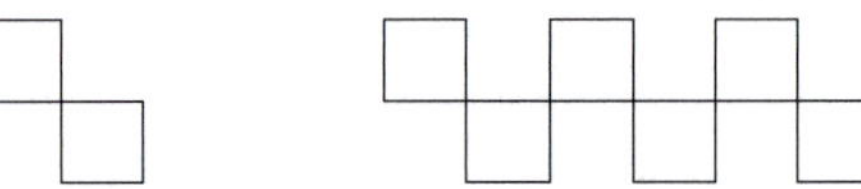
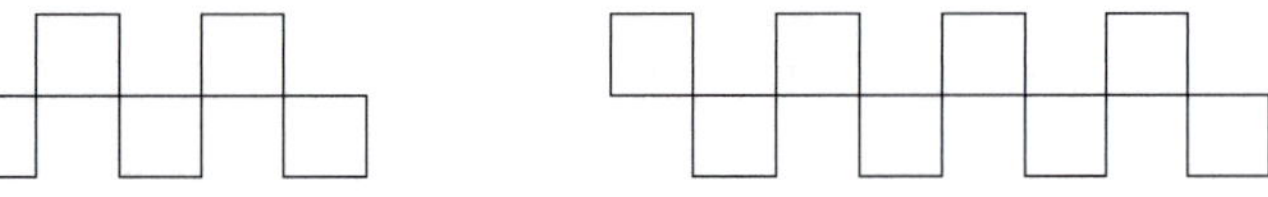
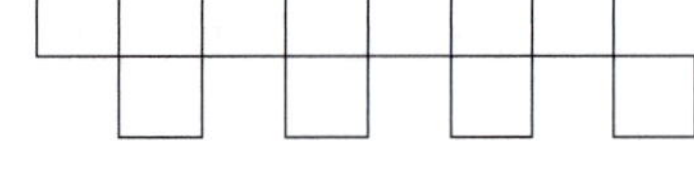

2 4 ________ ________

2 Number of popsicle sticks

________ ________ ________

3 Number of dots

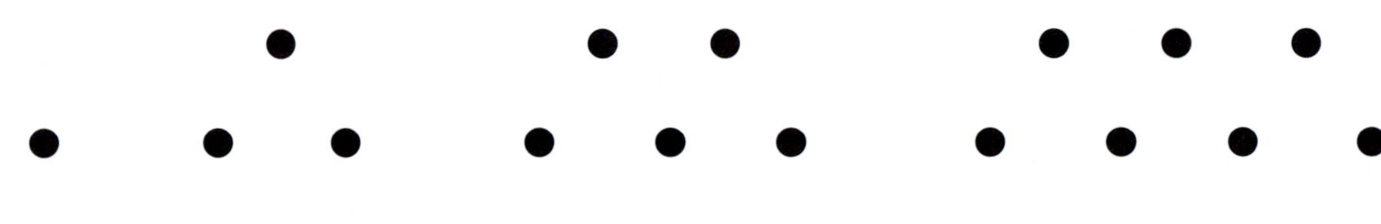

________ ________ ________ ________

4 Number of popsicle sticks

________ ________ ________

5 Number of popsicle sticks

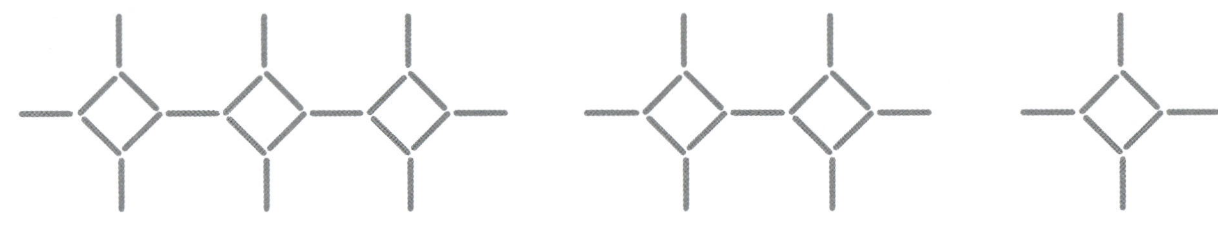

________ ________ ________

6 Number of triangles

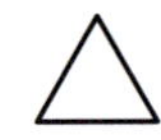
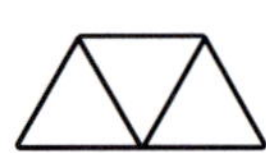
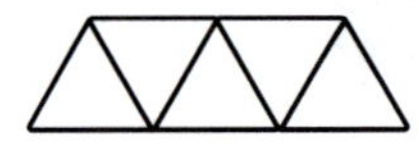
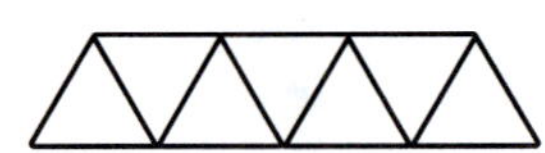

________ ________ ________ ________

ISBN: 9780170451505

Describing patterns

Fill in the gaps to complete the descriptions of these patterns.

1

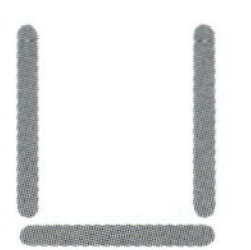  

I started with ______ **popsicle stick(s)** and then I added/~~subtracted~~ 2 each time.

2

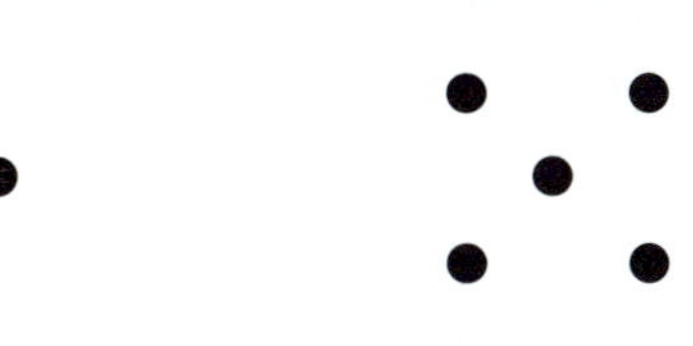 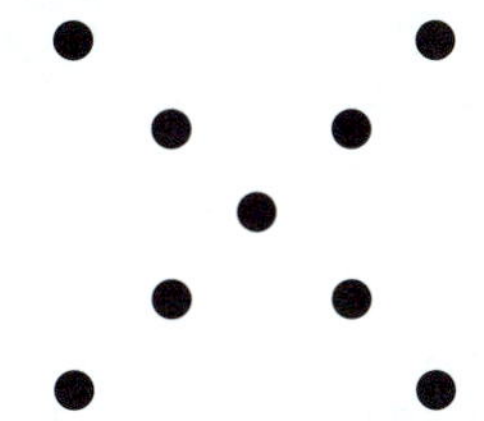

I started with ______ **dot(s)** and then I added/subtracted ______ each time.

3

I started with ______ **star(s)** and then I added/subtracted ______ each time.

4

I started with ______ **square(s)** ______________________________.

5

× × × ×　　　× × ×　　　× ×
× × × × × ×　　× × × × ×　　× × × ×

I started with ______ **cross(es)** ______________________________.

ISBN: 9780170451505

Challenge 1

There is a mistake in the final diagram for each of these sequences. Find the mistake and correct it.

1

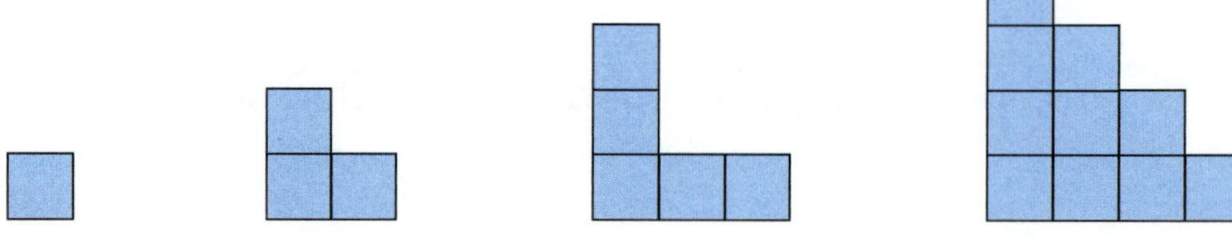

2

3

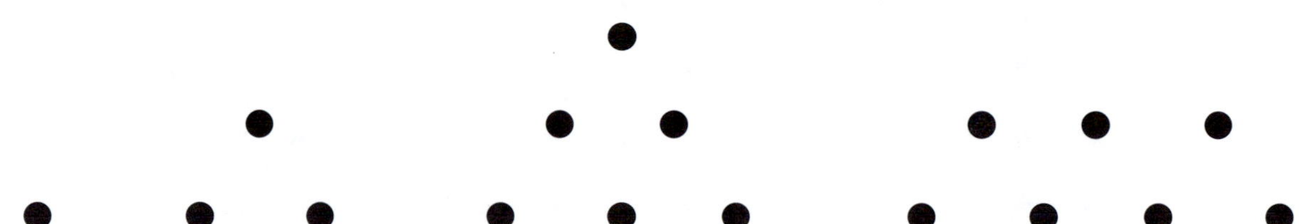

In these sequences, the mistake could be anywhere.

4

5 1, 3, 5, 8, 9, 11

6 72, 69, 66, 63, 61, 57

7 –2, 1, 5, 9, 13, 17

8 10, 3, –4, –10, –18, –25

 ISBN: 9780170451505

Patterns into tables

- Tables are a very useful way of showing patterns.
- We can use tables:
  1 to show us the pattern, and
  2 to predict, for instance, the number of popsicle sticks needed for bigger shapes **without having to draw them**.

Example:

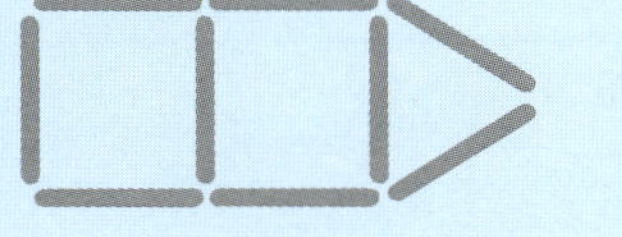
Shape 1

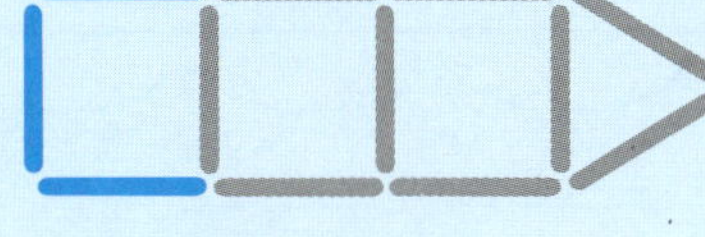
Shape 2

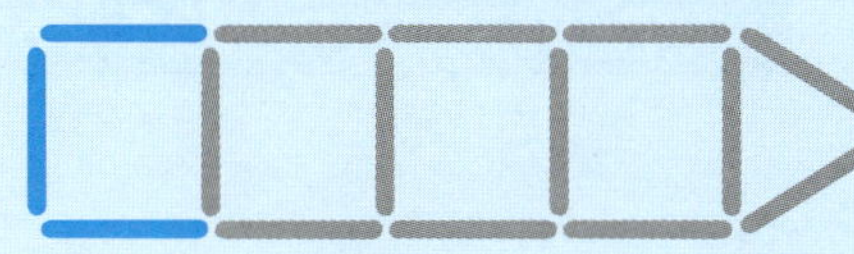
Shape 3

Shape number	Number of popsicle sticks
1	9
2	12
3	15
4	**18**
5	**21**
6	**24**

+ 3, + 3, + 3, + 3, + 3

To find the number in the next term, we need to **add 3**. You can see this in the shapes: **3 more** popsicle sticks were added each time.

Because we added 3 for the first row, we **keep adding 3** to complete the table — **without** having to draw the pictures.

Complete the tables for the shapes that have been drawn (the white boxes). Then, without drawing the next shapes, fill in the grey boxes.

1

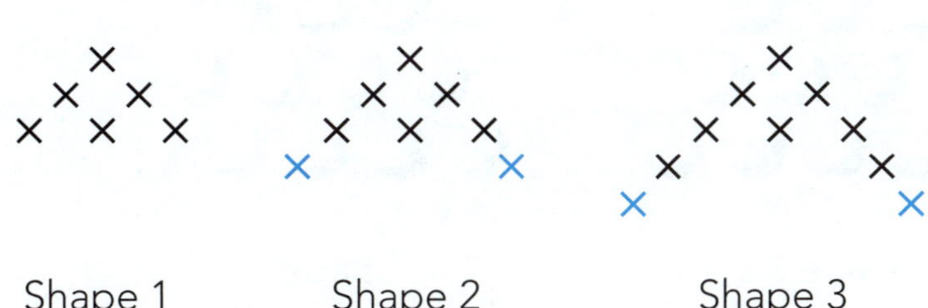
Shape 1 Shape 2 Shape 3

Shape number	Number of crosses
1	6
2	
3	
4	
5	
6	

2

Shape 1 Shape 2 Shape 3

Shape number	1	2	3	4	5	6
Number of popsicle sticks	26	21				

3

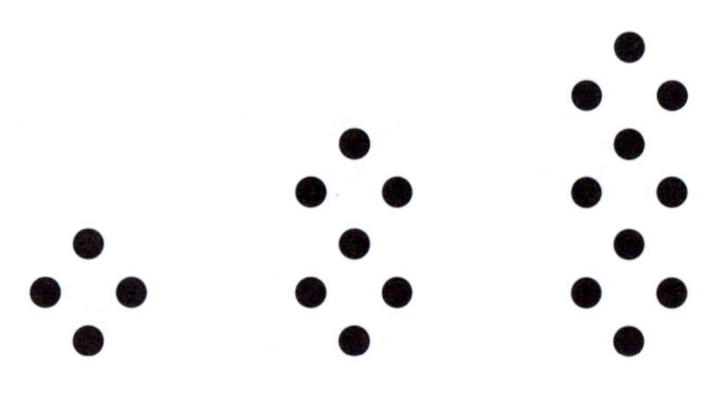

Shape 1 Shape 2 Shape 3

Shape number	Number of dots
1	4
2	
3	
4	
5	
6	

4

Shape 1 Shape 2 Shape 3

Shape number	1	2	3	4	5	6
Number of moons	15					

 ISBN: 9780170451505

Finding term 'zero'

- Not all patterns involve **shapes**. For example, the odd numbers between 10 and 20 form a pattern: 11, 13, 15, 17, 19.
- So from now on we will use the word '**term**' instead of 'shape'.
- It will also be very useful to find the **value of term 0**.

Example:

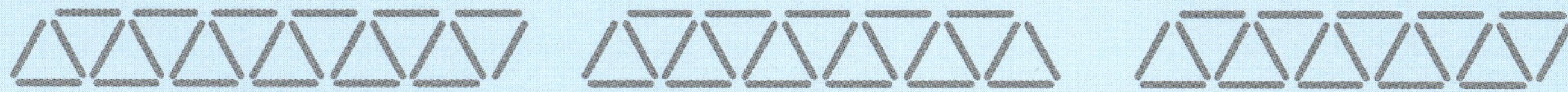

Term number	Number of popsicle sticks
0	**27**
1	25
2	23
3	21
4	19
5	17

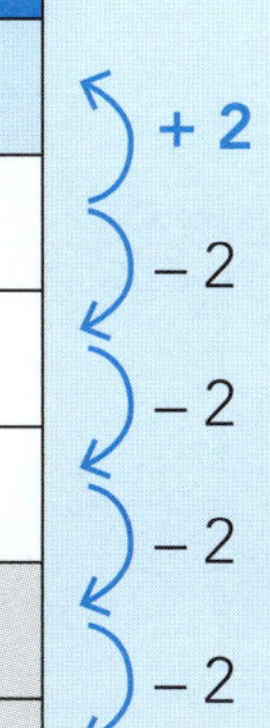

To find the bigger terms we **subtract 2**. So to find term zero we need to **add 2** to the first term.
25 **+ 2** =27

Complete the tables, including the value for term zero.

1

Term number	Number of squares
0	
1	6
2	
3	
4	
5	

2

× × × ×
× × × × × ×

× × ×
× × × × ×

× ×
× × × ×

Term number	0	1	2	3	4	5
Number of crosses				6		

ISBN: 9780170451505

3

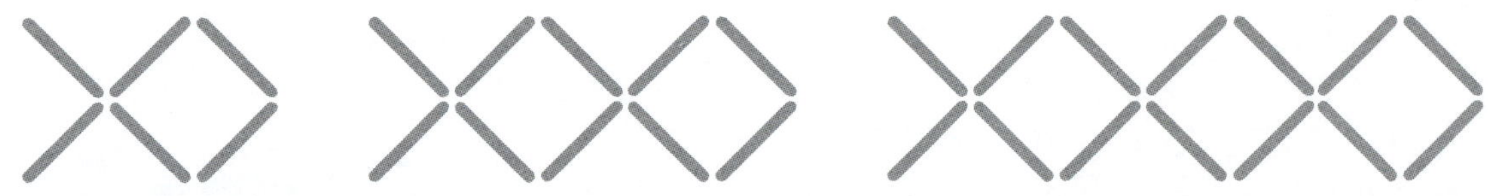

Term number	0	1	2	3	4	5	6
Number of popsicle sticks		6					

4

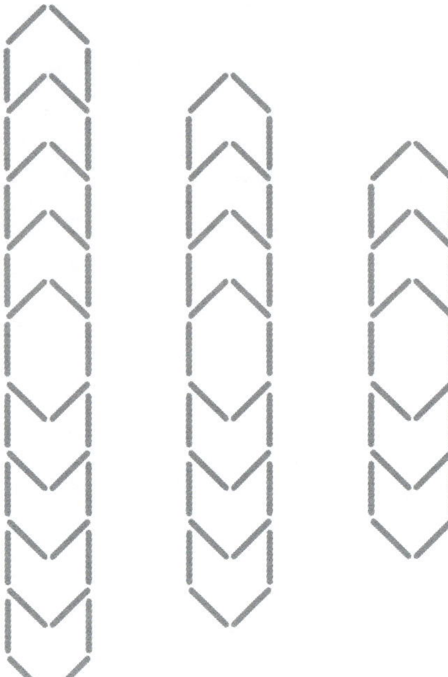

Term number	Number of popsicle sticks
0	
1	
2	
3	22
4	
5	

5

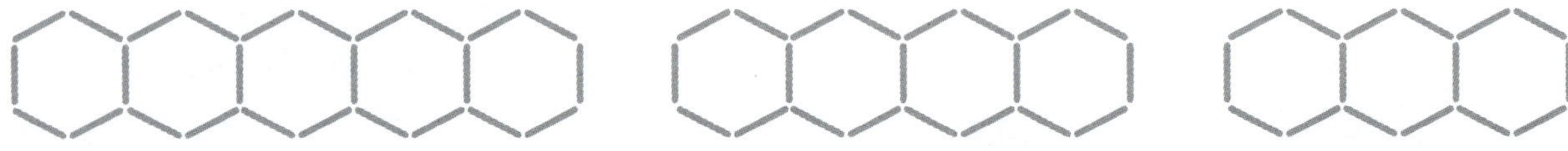

Term number	0	1	2	3	4	5	6
Number of popsicle sticks							

6

Term number	0	1	2	3	4	5	6
Number of popsicle sticks							

ISBN: 9780170451505

Finding the rule from a pattern

Patterns and tables

Think about the answers to the times tables:

Table	0	1	2	3	4	5	6	Change at each step of the table
x 3	0	3	6	9	12	15	18	+ 3
x 4	0	4	8	12	16	20	24	+ 4
x 5	0	5	10	15	20	25	30	+ 5

Now substitute the values of *n* into these expressions:

Value of n	0	1	2	3	4	5	6	Change at each step in the value of n
$4n$	0	4	8	12	16	20	24	+ 4
$4n + 3$	3	7	11	15	19	23	27	+ 4
$4n - 1$	–1	3	7	11	15	19	23	+ 4

So, if a pattern goes up in fours, then the equation will contain a $4n$.

Decreasing patterns

Table	0	1	2	3	4	5	6	Change at each step of the table
x (–3)	0	–3	–6	–9	–12	–15	–18	– 3
x (–4)	0	–4	–8	–12	–16	–20	–24	– 4
x (–5)	0	–5	–10	–15	–20	–25	–30	– 5

Now substitute the values of *n* into these expressions:

Value of n	0	1	2	3	4	5	6	Change at each step in the value of n
$-5n$	0	–5	–10	–15	–20	–25	–30	– 5
$-5n + 2$	2	–3	–8	–13	–18	–23	–28	– 5
$-5n - 4$	–4	–9	–14	–19	–24	–29	–34	– 5

So, if a pattern goes down in fives, then the equation will contain a $-5n$.

ISBN: 9780170451505

Increasing patterns

- It is useful to have a **mathematical rule** for patterns so we can work out, for instance, the value of the hundredth term **without** having to draw pictures or create a large table.
- Mathematical rules are easiest if they are written using **letters** (which we call **variables**) rather than words.

Examples:

1

Term number (n)	Number of popsicle sticks (P)
0	1
1	4
2	7
3	10
4	13
5	16

Differences: − 3 (from term 1 back to term 0), + 3, + 3, + 3, + 3

We could draw shape number 0: it would be just 1 popsicle stick.

Remember, do the **opposite** (−3) to find term 0.

In words: I started with 4 popsicle sticks and then I **added** 3 each time.

Find the rule: Number of Popsicle sticks = 3 x term number + 1

Tidy it up: $P = 3n + 1$

Use P to represent the number of Popsicle sticks.

Use n to represent the term number.

This is the number of popsicle sticks that you need for **term number 0.**

Word rule: The number of Popsicle sticks is calculated by multiplying the term number by **3** and then **adding 1**.

Use the rule: To find the number of popsicle sticks needed for the 100th shape:

$n = 100$ so $P = 3n + 1$

$= 3 \times 100 + 1$

$= 301$

We have found the number of popsicle sticks needed for the 100th shape **without** having to draw all the shapes or create a huge table.

ISBN: 9780170451505

2

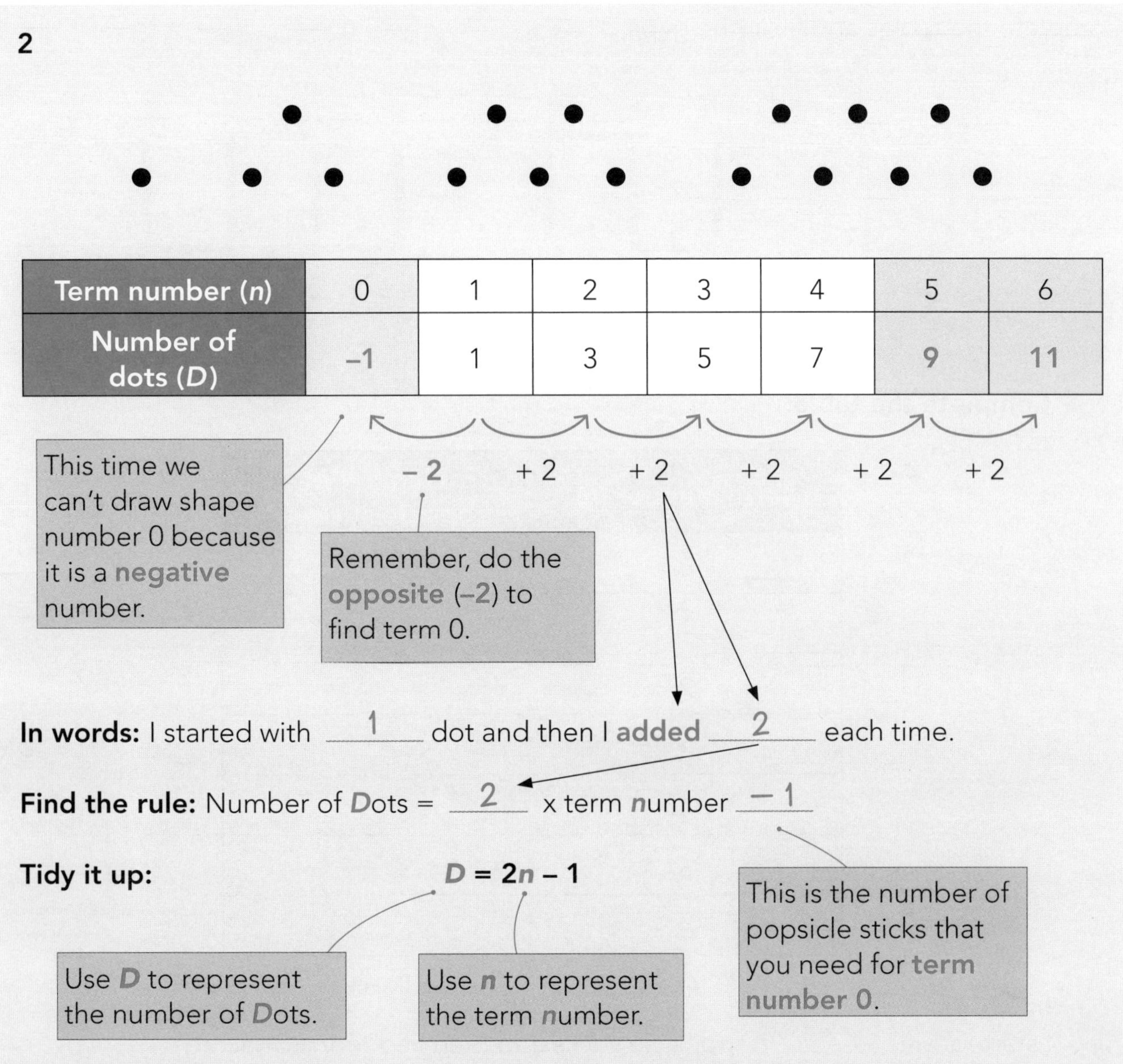

Term number (n)	0	1	2	3	4	5	6
Number of dots (D)	−1	1	3	5	7	9	11

In words: I started with 1 dot and then I **added** 2 each time.

Find the rule: Number of **D**ots = 2 x term ***n***umber − 1

Tidy it up: $D = 2n - 1$

Use ***D*** to represent the number of **D**ots.

Use ***n*** to represent the term ***n***umber.

This is the number of popsicle sticks that you need for **term number 0**.

Word rule: The number of **D**ots is calculated by multiplying the term ***n***umber by **2** and then **subtracting 1**.

Use the rule: To find the number of dots needed for the 100th shape:

$n = 100$ so $D = 2n - 1$

$= 2 \times 100 - 1$

$= 199$

Again, we have found the number of dots needed for the 100th shape **without** having to draw all the shapes or create a huge table.

Complete the tables and fill in the gaps.

1

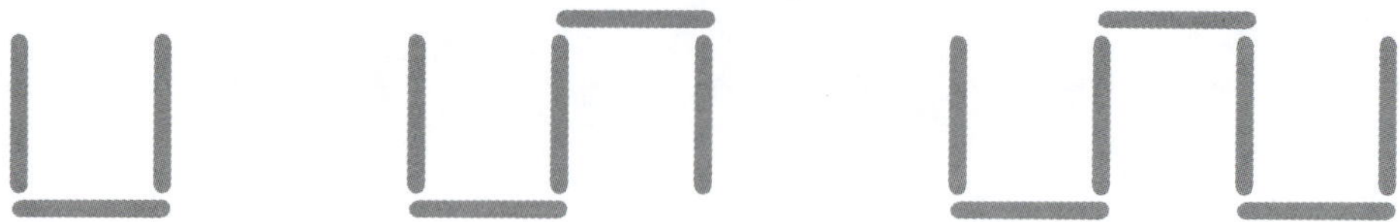

Complete the table:

Term number (n)	Number of popsicle sticks (P)
0	
1	3
2	
3	
4	
5	
6	

In words:
I started with ________ popsicle sticks and then I added/~~subtracted~~ ________ each time.

Find the mathematical rule:

Number of popsicle sticks = ________ x term number + ________

Tidy it up: $P =$ ________ $n +$ ________

Word rule:
The number of popsicle sticks is calculated by multiplying the term number by ________ and then adding/subtracting ________.

Use the rule:
How many popsicle sticks would be needed for the 40th shape?

$P =$ ______ x 40 ________

= ________

 ISBN: 9780170451505

2

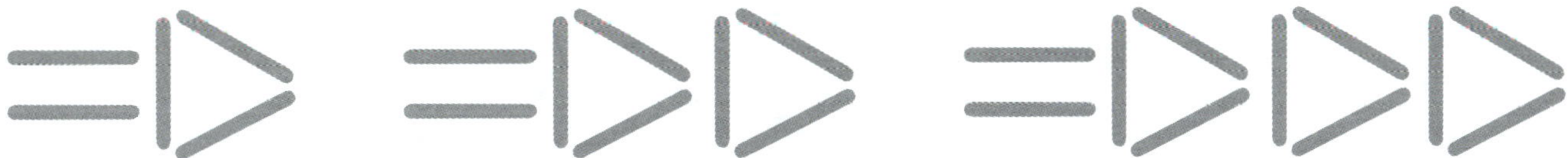

Complete the table:

Term number (n)	0	1	2	3	4	5	6
Number of popsicle sticks (P)		5					

In words:
I started with ________ popsicle sticks and then I added/subtracted ________ each time.

Find the mathematical rule:

Number of popsicle sticks = ________ x term number ________

Tidy it up: P = ________n ________

Word rule:
The number of popsicle sticks is calculated by multiplying the term number by ________ and then adding/subtracting ________.

Use the rule:
How many popsicle sticks would be needed for the 50th shape?

P = ________________

= ________________

ISBN: 9780170451505

3

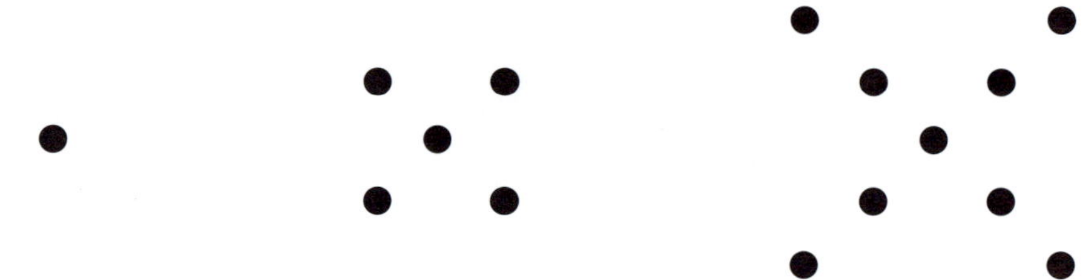

Complete the table:

Term number (n)	Number of dots (D)
0	
1	
2	
3	
4	
5	
6	

In words:
I started with ____________ dots and then I added/subtracted ____________ each time.

Find the mathematical rule:

Number of dots = ____________ x term number ____________

Tidy it up: D = ____________ n ____________

Word rule:
The number of dots is calculated by multiplying the term number by ____________ and then adding/subtracting ____________.

Use the rule:
How many dots would be needed for the 40th shape?

D = ____________________

= ____________________

 ISBN: 9780170451505

4

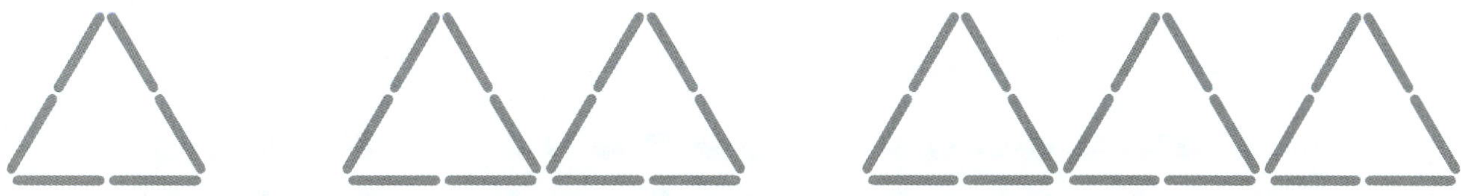

Complete the table:

Term number (n)	0	1	2	3	4	5	6
Number of popsicle sticks (P)							

In words:
I started with ________ popsicle sticks and then I added/subtracted ________ each time.

Find the mathematical rule:

Number of popsicle sticks = ________ x term number ________

Tidy it up: P = ____________________

Word rule:
The number of popsicle sticks is calculated by multiplying the term number by ________ and then adding/subtracting ________.

Use the rule:
How many popsicle sticks would be needed for the 50th shape?

P = ____________________

= ____________________

Decreasing patterns

- Some patterns **reduce** in size as the term number increases.

Examples:

1

Term number (n)	Number of dots (D)
0	**16**
1	14
2	12
3	10
4	**8**
5	**6**

+ 2, − 2, − 2, − 2, − 2

We could draw term number 0: it would have 16 dots:

●●●●●●●●
●●●●●●●●

Remember, do the **opposite** (**+ 2**) to find term 0.

In words: I started with 14 popsicle sticks and then I **subtracted** 2 each time.

Find the rule: Number of **D**ots = −2 x term **n**umber + 16

Notice that for reducing patterns, we need to multiply the term number by a **negative** number.

This is the number of popsicle sticks that you need for **term number 0**.

Tidy it up: $D = -2n + 16$

Word rule: The number of **D**ots is calculated by multiplying the term **n**umber by **−2** and then **adding 16**.

Use the rule: To find the number of dots needed for the 7th shape:

$n = 7$ so $D = -2n + 16$
$= -2 \times 7 + 16$
$= 2$

Again, we have found the number of dots needed for the 7th shape **without** having to draw all the shapes or create a larger table.

 ISBN: 9780170451505

Complete the tables and fill in the gaps.

1

Complete the table:

Term number (n)	Number of squares (S)
0	
1	
2	
3	
4	
5	

In words:
I started with ________ squares and then I added/subtracted ________ each time.

Find the mathematical rule:

Number of squares = ________ x term number ________

Tidy it up: S = ________n ________

Word rule:
The number of squares is calculated by multiplying the term number by ________ and then adding/subtracting ________.

Use the rule:
How many squares would be needed for the 6th shape?

S = ________________

= ________________

ISBN: 9780170451505

2

Complete the table:

Term number (n)	0	1	2	3	4	5
Number of popsicle sticks (P)						

In words:
I started with ________ popsicle sticks and then I added/subtracted ________ each time.

Find the mathematical rule:

Number of popsicle sticks = ________ x term number ________

Tidy it up: P = ________________

Word rule:
The number of popsicle sticks is calculated by multiplying the term number by ________ and then adding/subtracting ________.

Use the rule:
How many popsicle sticks would be needed for the 6th shape?

P = ________________

= ________________

 ISBN: 9780170451505

3

Complete the table:

Term number (n)	Number of snowflakes (S)
0	
1	
2	
3	
4	
5	

In words:
I started with ____________ snowflakes and then I added/subtracted ____________ each time.

Find the mathematical rule:

Number of snowflakes = ____________ x term number ____________

Tidy it up: S = ____________________________

Word rule:
The number of snowflakes is calculated by multiplying the term number by ____________ and then adding/subtracting ____________.

Use the rule:
How many snowflakes would be needed for the 7th shape?

S = ____________________________

= ____________________________

ISBN: 9780170451505

4

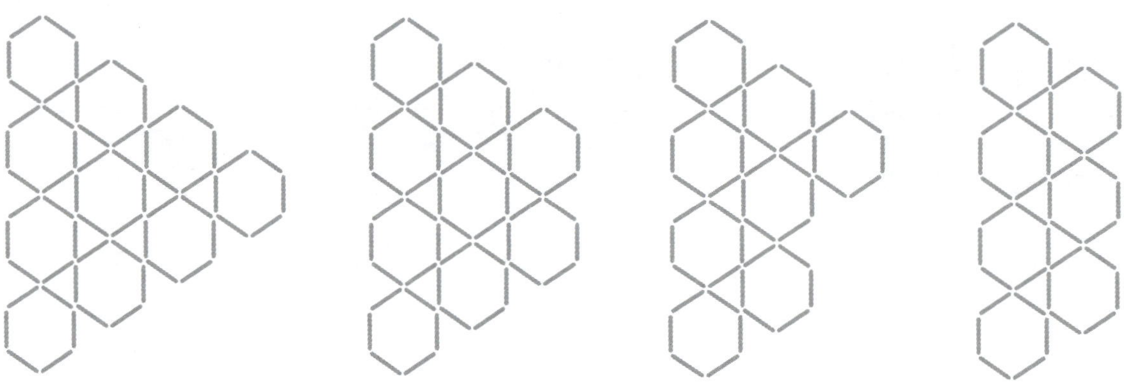

Complete the table:

Hint: Just count the hexagons.

Term number (n)	0	1	2	3	4	5
Number of popsicle sticks (P)						

In words:
I started with ________ popsicle sticks and then I added/subtracted ________ each time.

Find the mathematical rule:

Number of popsicle sticks = ________ x term number ________

Tidy it up: $P =$ ________________

Word rule:
The number of popsicle sticks is calculated by multiplying the term number by ________ and then adding/subtracting ________.

Use the rule:
How many popsicle sticks would be needed for the 10th shape?

$P =$ ________________

$=$ ________________

 ISBN: 9780170451505

Challenge 2

Count the popsicle sticks needed to create these patterns, and match each pattern to its rule.

• • $P = -4n + 18$

• • $P = 4n + 2$

• • $P = 4n$

• • $P = -5n + 31$

• • $P = 3n + 4$

• • $P = -5n + 28$

• • $P = 7n + 5$

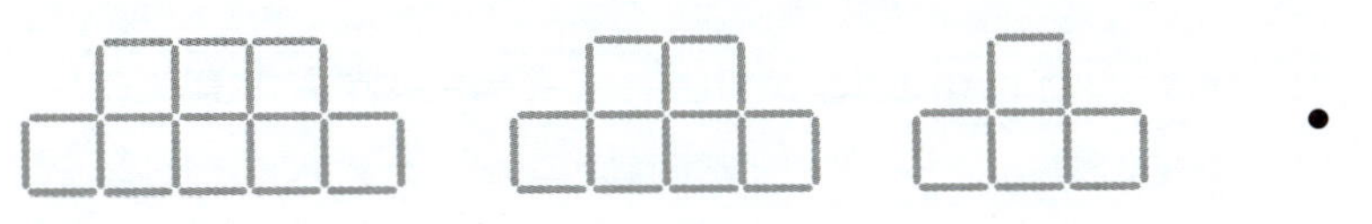

• • $P = 5n - 1$

Finding the rule from a table

- As we said earlier, not all patterns involve shapes: patterns are sometimes **described**.
- In these cases we need to need to use the description to draw up a table.

Examples:

1 The odd numbers that are bigger than 2.

Draw up a table:

Term number (n)	Value of term (T)
0	1
1	3
2	5
3	7
4	9
5	11

Find the rule:

The pattern **increases** by **2** each time.

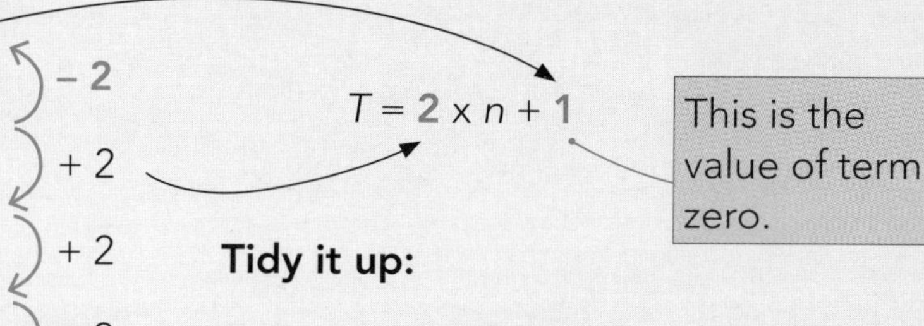

Tidy it up:

$$T = 2n + 1$$

Use the rule:

Find the 50th odd number that is bigger than 2.

$T = 2n + 1$
$= 2 \times 50 + 1$
$= 101$

2 Decreasing multiples of 4 starting from 100.

Draw up a table:

Term number (n)	0	1	2	3	4	5	6
Value of term (T)	104	100	96	92	88	84	80

+ 4   – 4   – 4   – 4   – 4   – 4

The pattern **decreases** by **4** each time.

Find the rule: $T = -4 \times n + 104$

Tidy it up: $T = -4n + 104$

Use the rule:

Starting from 100 and decreasing, find the 20th multiple of 4.

$T = -4n + 104$
$= -4 \times 20 + 104$
$= 24$

ISBN: 9780170451505

Complete the tables and find the rules for the following.

1 Increasing even numbers that are bigger than 5.

Term number (n)	Value of term (T)
0	
1	6
2	
3	
4	

Rule:

$T =$ ______ $n +$ ______

______

2 Start at 6, and add 3 each time.

Term number (n)	0	1	2	3	4	5
Value of term (T)						

______

Rule:

$T =$ ______ $n +$ ______

3 Increasing multiples of 4 that are bigger than 1.

Term number (n)	0	1	2	3	4	5
Value of term (T)						

Rule:

$T =$ ______ $n +$ ______

4 Increasing multiples of 10 that are bigger than 9.

Term number (n)	Value of term (T)
0	
1	
2	
3	
4	

Rule:

$T =$ ______

5 Decreasing odd numbers that are smaller than 20.

Term number (n)	0	1	2	3	4	5
Value of term (T)						

Rule:

$T =$ ______

ISBN: 9780170451505

6 Start at 50 and go down 5 each time.

Term number (n)	Value of term (T)
0	
1	
2	
3	
4	

Rule:

$T =$ ______________

7 Start at 4 and go up in 6s.

Term number (n)	1	2	3	4	5
Value of term (T)					

Rule:

$T =$ ______________

8 Start at 2 and subtract 3 each time.

Term number (n)	1	2	3	4	5
Value of term (T)					

Rule:

$T =$ ______________

9 Decreasing multiples of 4, starting at 100.

Term number (n)	Value of term (T)
1	
2	
3	
4	

Rule:

$T =$ ______________

10 Decreasing even numbers, starting at –4.

Term number (n)	1	2	3	4	5
Value of term (T)					

Rule:

$T =$ ______________

ISBN: 9780170451505

Challenge 3

Find the missing pictures or numbers for these sequences.

1

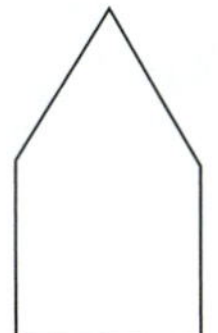

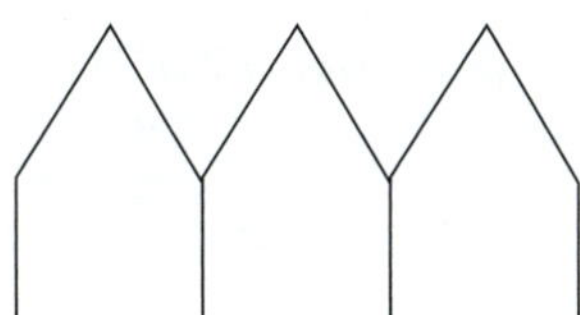

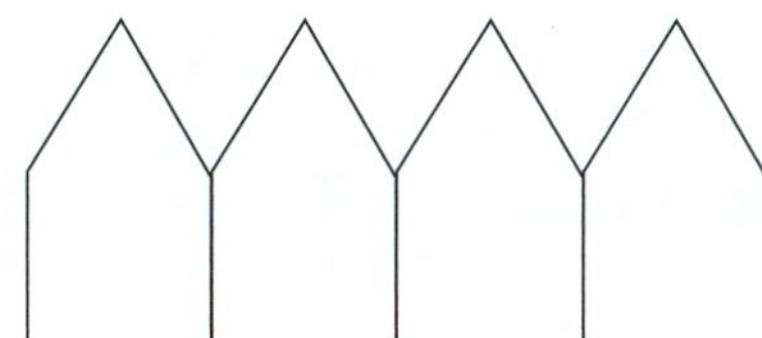

2

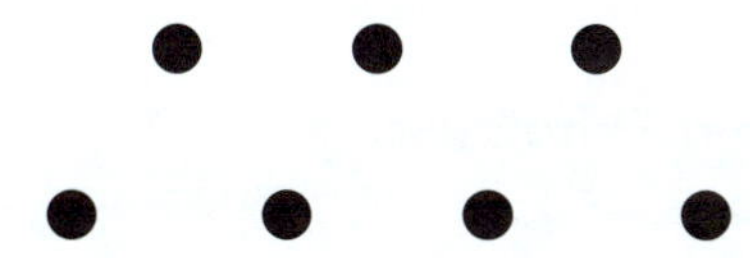

3

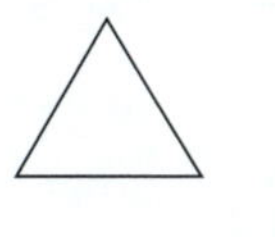

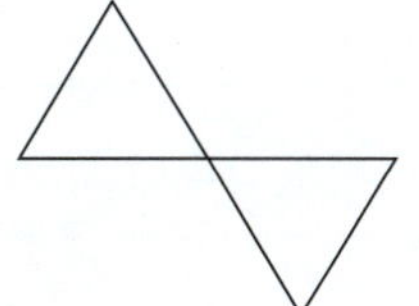

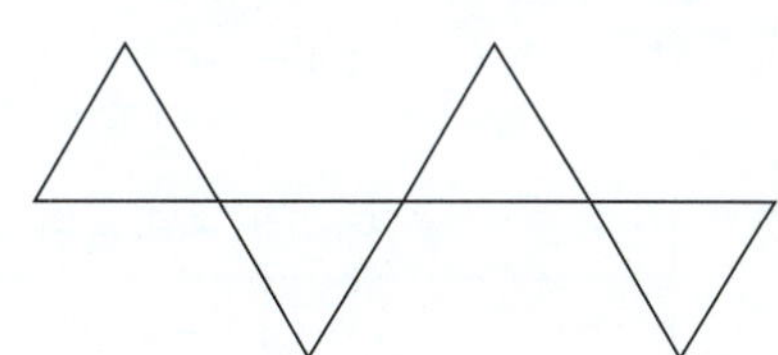

4 6, 12, 18, ______, 30, 36

5 7, 9, 11, 13, ______, 17, ______

6 12, ______, 36, ______, 60, 72

7 2, 5, ______, ______, 14, ______

8 ______, 34, 40, ______, 52, ______

9 ______, 75, ______, 57, ______, 39, ______

10 81, ______, ______, ______, ______, 51

ISBN: 9780170451505

Finding the rule from a list

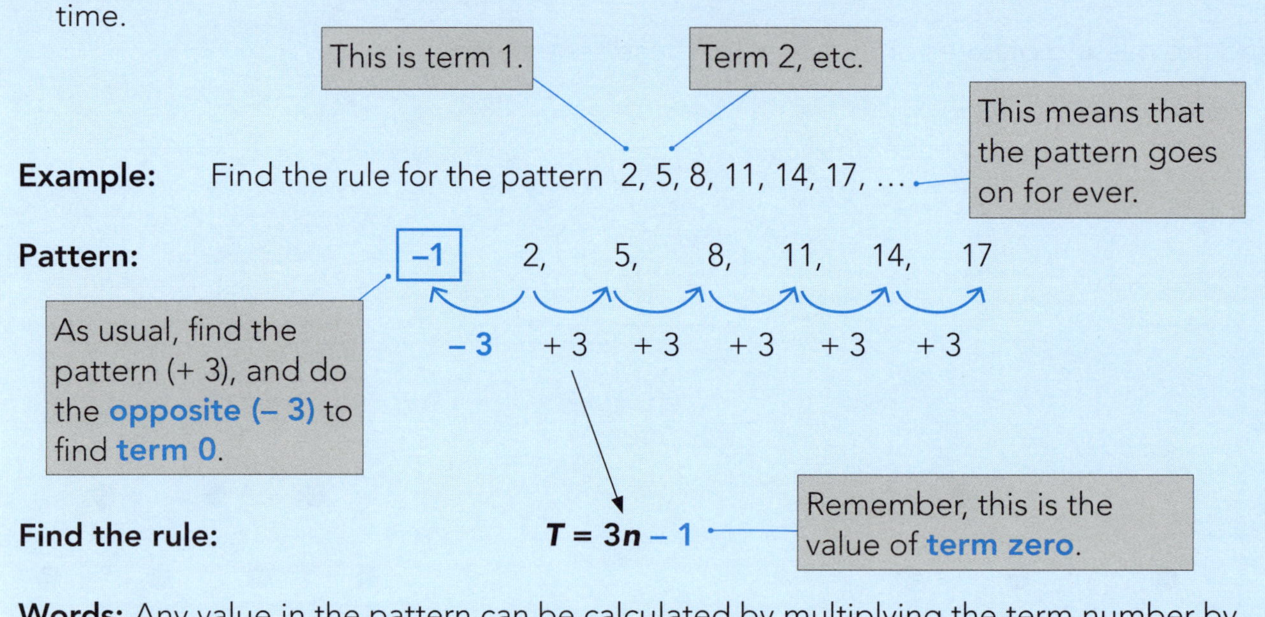

Words: Any value in the pattern can be calculated by multiplying the term number by **3** and then subtracting **1**.

Pick the correct rule from the options below.

1

8, 14, 20, 26, 32, …	
$T = 2n + 6$	$T = 6n + 6$
$T = 6n + 2$	$T = 6n + 8$

2

11, 14, 17, 20, 23, …	
$T = 3n + 8$	$T = 3n + 11$
$T = 8n + 8$	$T = 8n + 3$

Check your answers before you continue.

3

3, 7, 11, 15, 19, …	
$T = 4n + 3$	$T = 4n - 3$
$T = -4n - 1$	$T = 4n - 1$

4

16, 14, 12, 10, 8, …	
$T = 2n + 18$	$T = -2n + 16$
$T = -2n + 18$	$T = 18n - 2$

5

30, 25, 20, 15, 10, …	
$T = 5n + 35$	$T = -5n + 35$
$T = 35n - 5$	$T = -5n + 30$

6

0, 10, 20, 30, 40, …	
$T = 10n$	$T = 10n - 10$
$T = -10n$	$T = -10n - 10$

7

6, 3, 0, –3, –6, …	
$T = -3n + 6$	$T = -3n + 9$
$T = 3n + 6$	$T = -3n - 9$

8

–12, 0, 12, 24, 36, …	
$T = 12n + 24$	$T = -12n + 24$
$T = 12n - 24$	$T = 12n - 12$

ISBN: 9780170451505

Complete the statements and find the rule for each of these.

9 4, 7, 10, 13, …

___ ___ ___ ___

Rule:

$T =$ _____ $n +$ _____

I started with _____ and then I added/subtracted _____ each time.
Term zero will be _____.

10 1, 3, 5, 7, 9, …

Rule:

$T =$ _____ $n -$ _____

I started with _____ and then I added/subtracted _____ each time.
Term zero will be _____.

11 24, 20, 16, 12, 8, …

Rule:

$T =$ _____ $n +$ _____

I started with _____ and then I added/subtracted _____ each time.
Term zero will be _____.

12 –5, 0, 5, 10, 15, …

Rule:

$T =$ _____

I started with _____ and then I added/subtracted _____ each time.
Term zero will be _____.

13 20, 14, 8, 2, –4, …

Rule:

$T =$ _____

I started with _____ and then I added/subtracted _____ each time.
Term zero will be _____.

14 12, 22, 32, 42, 52, …

Rule:

$T =$ _____

15 6, 4, 2, 0, –2, …

Rule:

$T =$ _____

16 13, 16, 19, 22, 25, …

Rule:

$T =$ _____

17 9, 18, 27, 36, 45, …

Rule:

$T =$ _____

ISBN: 9780170451505

Finding a term from a rule

- There are many practical situations where it's useful to find the value of a term.
- From now on, we will use the term **sequence**, rather than pattern.
- A sequence is a list of numbers or objects **in a particular order**.
- Sometimes sequences occur in everyday situations.
- We can use the rule to find the value of any term in the sequence.
- This is much faster than writing a list.

Examples:

1 Find the 60th term for the sequence that has the rule $T = 3n + 2$.

$T = 3n + 2$
$= 3 \times \mathbf{60} + 2$
$= 182$

Substitute **60** for ***n***.

2 Find the 80th term for the sequence that has the rule $T = -2n + 10$.

$T = -2n + 10$
$= -2 \times \mathbf{80} + 10$
$= -150$

Use the rules to calculate the terms.

1 50th term $T = 2n + 1$
= ____________
= ________

2 20th term $T = 5n + 15$
= ____________
= ________

3 50th term $T = 20n + 100$
= ____________
= ________

4 50th term $T = -2n + 6$
= ____________
= ________

5 8th term $T = 6n - 10$
= ____________
= ________

6 10th term $T = 3n - 40$
= ____________
= ________

7 10th term $T = -4n + 50$
= ____________
= ________

8 5th term $T = -7n + 15$
= ____________
= ________

9 4th term $T = -3n - 1$
= ____________
= ________

10 8th term $T = -10n - 40$
= ____________
= ________

ISBN: 9780170451505

Finding the sequence from the rule

- We can use the rule to find the sequence.

Examples:

1 Rule: $T = 3n + 2$

Use substitution, along with the pattern, to find the terms:

Term number (n)	1	2	3	4	5
Calculations	$T = 3 \times 1 + 2$				$T = 3 \times 5 + 2$
Value of term (T)	5	8	11	14	17

+ 3 + 3 + 3 + 3

a What is the first term? $T = 3 \times 1 + 2 = 5$

b What does the sequence go up or down by? + 3

c Check term 5. $T = 3 \times 5 + 2 = 17$ ✓

Tidy it up: 5, 8, 11, 14, 17

2 Rule: $T = -2n + 8$

Term number (n)	Calculations	Value of term (T)
1	$-2 \times 1 + 8$	6
2		4
3		2
4		0
5	$-2 \times 5 + 8$	−2

− 2 − 2 − 2 − 2

a What is the first term? $T = -2 \times 1 + 8 = 6$

b What does the sequence go up or down by? − 2

c Check term 5. $T = -2 \times 5 + 8 = -2$ ✓

Tidy it up: 6, 4, 2, 0, −2

Find the first five terms of these sequences using the rule.

1 $T = 5n + 2$

Term number (n)	1	2	3	4	5
Calculations	$5 \times 1 + 2$				
Value of term (T)					

Sequence: ______________________________

ISBN: 9780170451505

2 $T = -3n + 10$

Term number (n)	1	2	3	4	5
Calculations					
Value of term (T)					

Sequence: ______________________________

3 $T = 6n - 1$

Term number (n)	Calculations	Value of term (T)
1		
2		
3		
4		
5		

Sequence: ______________________________

4 $T = 4n - 11$

Term number (n)	1	2	3	4	5
Calculations					
Value of term (T)					

Sequence: ______________________________

5 $T = 10n - 100$

Term number (n)	1	2	3	4	5
Calculations					
Value of term (T)					

Sequence: ______________________________

6 $T = -5n - 2$

Term number (n)	1	2	3	4	5
Calculations					
Value of term (T)					

Sequence: ______________________________

ISBN: 9780170451505

Applications

Example: Emily has started a part-time job, so she is able to deposit $20 into her savings account at the end of each week. Let T represent the total she has in her bank account, and n represent the number of weeks since she started her job.

a At the end of her first week she has $185 in the account. How much did she start with?

She started with $185 – $20 = $165

b Which term does $165 represent?

Term 0

c Write a list for the sequence showing the amount she has in her savings account, starting from $n = 1$.

$185, $205, $225, $245, …

+ 20

d Write the rule.

$T = 20n + 165$

e Use the rule to calculate how much she will have in her account after 9 weeks.

$T = 20 \times 9 + 165$

$= \$345$

Answer the following questions.

1 Alex has started a part-time job, so he is able to deposit $15 into his savings account at the end of each week. Let T represent the total he has in his bank account, and n represent the number of weeks since he started his job.

a At the end of his first week he has $75 in the account. How much did he start with?

He started with $ ____________

b What term does his starting amount represent?

____

c Write a list for the sequence showing the amount he has in his bank account, starting at $n = 1$.

________, ________, ________, …

d Write the rule.

$T =$ ____________

e Use the rule to calculate how much he will have in his account after 11 weeks.

$T =$ ____________

$=$ ____________

ISBN: 9780170451505

2 Hiring a Crumbling Kayak costs \$10, plus \$11 for each hour it is used. Let T be the total cost of hiring a kayak for n hours.

a Write a list for the sequence starting at $n = 1$. ______, ______, ______, …

b Calculate term zero. Term zero = ______

c Write the rule. $T =$ ______

d Use the rule to calculate how much it will cost to hire a kayak for 8 hours. $T =$ ______

= ______

3 Hiring a Crafty Canoe costs \$16 , plus \$8 for each hour it is used. Let T be the total cost of hiring a canoe for n hours.

a Write a list for the sequence starting at $n = 1$. ______, ______, ______, …

b Calculate term zero. Term zero = ______

c Write the rule. $T =$ ______

d Use the rule to calculate how much it will cost to hire a canoe for 9 hours. $T =$ ______

= ______

4 Grandma has 38 lollipops in her jar. She gives one each to her three grandchildren when they visit each Saturday. Let T be the total number of lollipops in the jar after n visits.

a Write a list for the sequence starting at $n = 1$. ______, ______, ______, …

b Calculate term zero. Term zero = ______

c Write the rule. $T =$ ______

d Use the rule to calculate how many lollipops she will have after 11 visits. $T =$ ______

= ______

ISBN: 9780170451505

5 Tama buys a phone which costs $299. In addition it will cost him $35 per month. Let T represent the total amount he has paid for the phone and its use, and n represent the number of months he has had it.

a Write a list for the sequence starting at $n = 1$. ______, ______, ______, …

b Calculate term zero. Term zero = ______

c Write the rule. $T =$ ______

d Use the rule to calculate how much it will have cost him after he has owned it for 12 months. $T =$ ______

= ______

6 Eli is building a set of steps. Each step needs 8 screws, and he has bought a pack of 100 screws. Let T be the number of screws remaining after he has built n steps.

a Write a list for the sequence starting at $n = 1$. ______, ______, ______, …

b Calculate term zero. Term zero = ______

c Write the rule. $T =$ ______

d Use the rule to calculate the number of screws remaining after he has built 9 steps. $T =$ ______

= ______

7 After a power cut, the temperature in a freezer rises to –3°C. Then the power comes back on, the temperature drops at a steady rate of 2°C per hour. Let T be the temperature in the freezer, and n be the number of hours since the power came back on.

a Write a list for the sequence starting at $n = 1$. ______, ______, ______, …

b Calculate term zero. Term zero = ______

c Write the rule. $T =$ ______

d Use the rule to calculate the temperature in the freezer 8 hours after the power comes back on. $T =$ ______

= ______

Challenge 4

Each row and column of these grids contains a linear sequence. Fill in the gaps.

1

2			8
5		15	
	16		
		33	

2

16		18	
	35		
		72	
43			

3

	9		5
		27	
35			

4

		30	
	7		
8			26
		15	

5

	5		
7		11	
			9
	17		

6

			0
		17	
–21			99

7

–1		21	
		–6	–1

8

–11			
	0		
46			–116

ISBN: 9780170451505

Cross-number

The answers to the clues below are the first terms of a sequence.

1			2 11		3	27		4
		5				6		
7					8			
9			10		11			12
		13						
14					15			

Across		Down	
1	$3n - 1$	**1**	$6n - 4$
3	$4n + 19$	**2**	$2n + 9$
5	$5n + 5$	**3**	$n + 22$
7	$-n + 21$	**4**	$-n + 36$
8	$2n + 24$	**5**	$8n + 2$
9	$2n + 28$	**6**	$-2n + 32$
11	$4n + 16$	**9**	$n + 29$
13	$-5n + 47$	**10**	$n + 35$
14	$2n + 31$	**11**	$7n + 13$
15	$-5n + 46$	**12**	$-2n + 34$

ISBN: 9780170451505

Challenge 5

1 Find and highlight the 14 sequences in this number grid. The sequences may be horizontal, vertical or diagonal, and each sequence is five terms long.

3	5	3	7	11	15	19	23	10	15
7	8	8	9	14	16	18	19	20	20
10	11	13	11	17	17	16	25	30	2
1	15	8	18	20	40	30	20	10	0
4	5	3	7	23	35	50	20	10	2
2	4	9	18	16	14	19	12	8	4
21	17	12	13	14	18	16	14	12	6
15	12	10	20	17	22	20	19	17	8
9	7	11	16	28	26	24	22	20	17
4	8	12	15	18	30	27	24	24	21

2 Fill in the gaps so that you complete these patterns.

a 22, 24, 28, 34, 42, 52, ________, ________, ________

b 70, 68, 65, 61, 56, 50, ________, ________, ________

c 3, 4, 8, 17, 33, 58, ________, ________, ________

 ISBN: 9780170451505

Graphs

Plotting points

Positive coordinates

- A positive **x** coordinate tells you how far to move to the **right**.
- A positive **y** coordinate tells you how far to move **up**.
- Coordinates are written in brackets, and in **alphabetical order**: **(x, y)**.
- We use the word 'ax**i**s' for one axis, and the word 'ax**e**s' for more than one.

Consider the point **A** on the graph.
It would be written as **(3, 4)**.

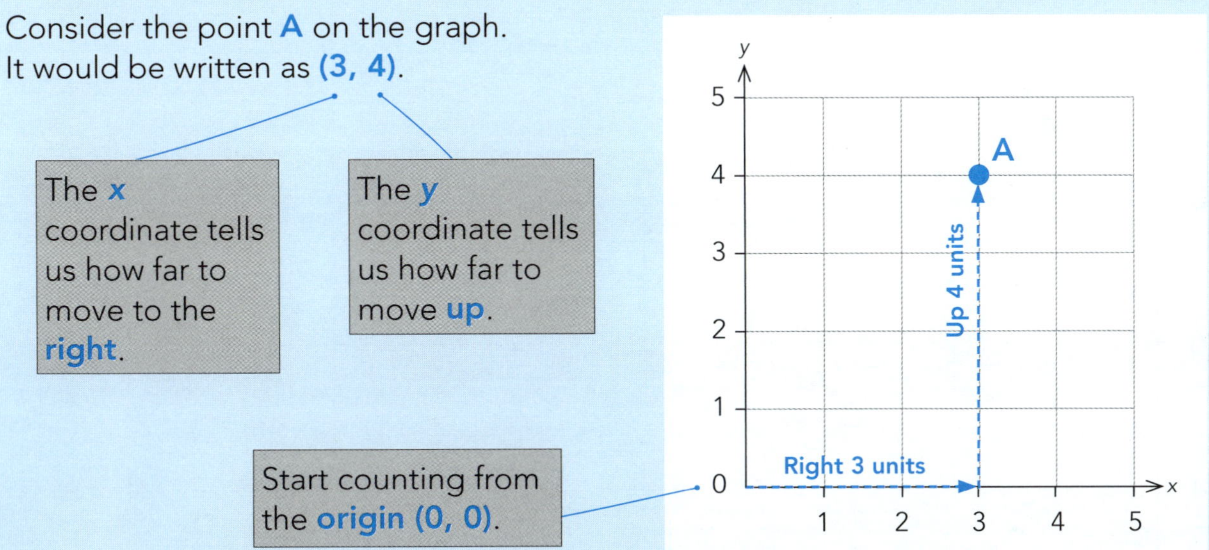

Sometimes not all the numbers are put on the axes.

Examples:

Only the **even** numbers are on these axes. Point **B** is at **(7, 9)**.

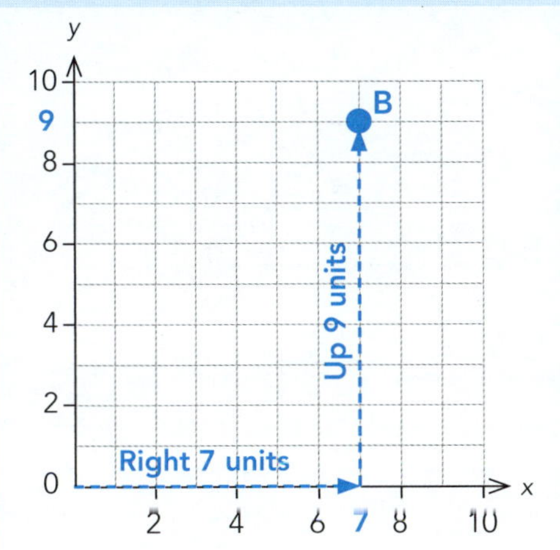

Only **multiples of 5** are on these axes. Point **C** is at **(11, 12)**.

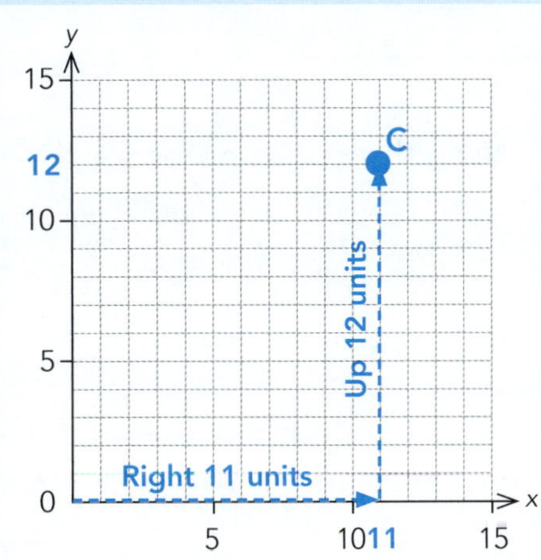

ISBN: 9780170451505

Answer the following questions.

1 **a** Add the missing numbers to these axes.

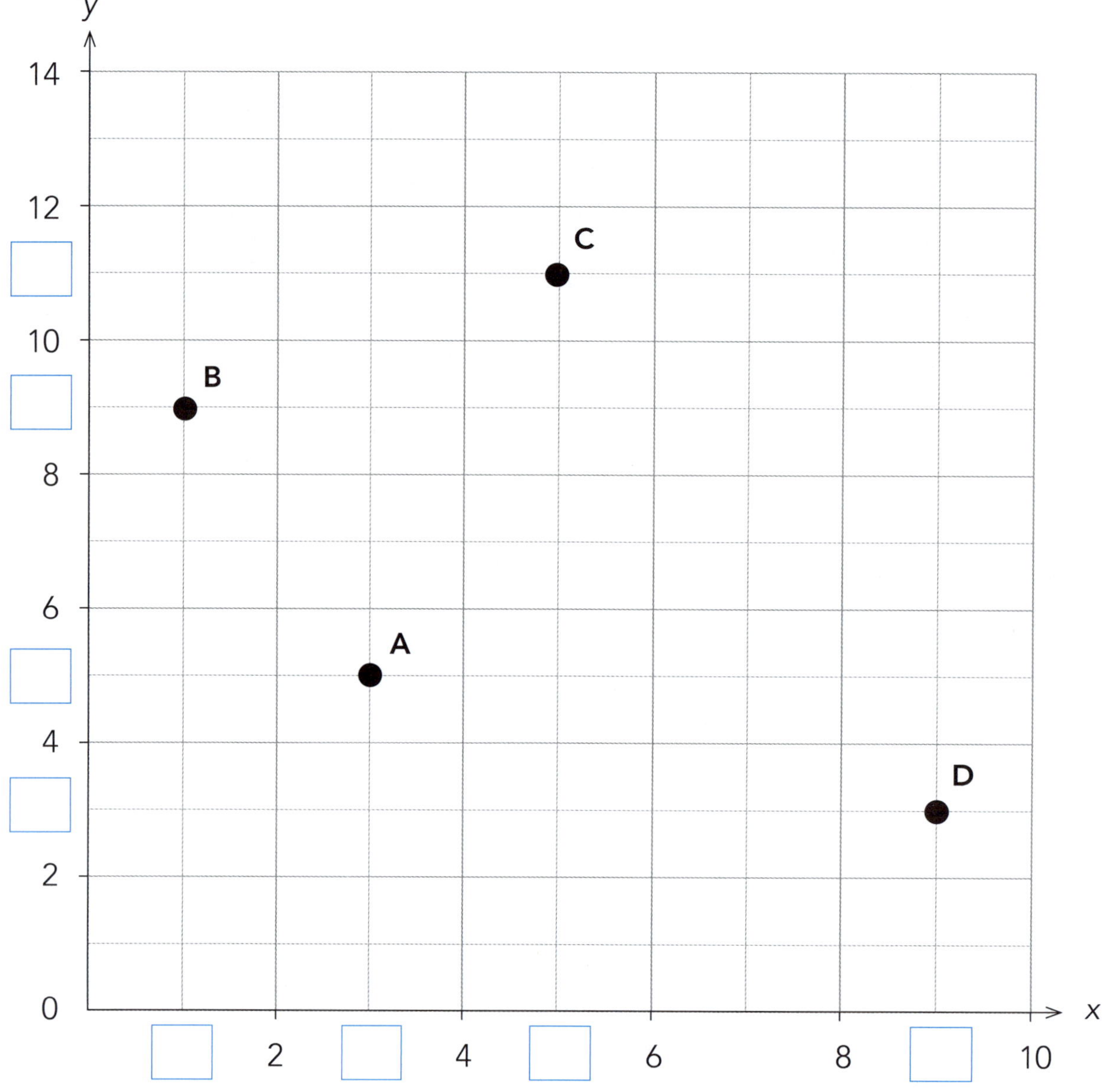

b Write the coordinates for the points A to D.

A (______, ______) B (______, ______)

C (______, ______) D (______, ______)

c Plot and label these points on the axes.

E (7,5) F (1, 1) J (3, 13) K (5, 3)

ISBN: 9780170451505

2 **a** Add the missing numbers to these axes.

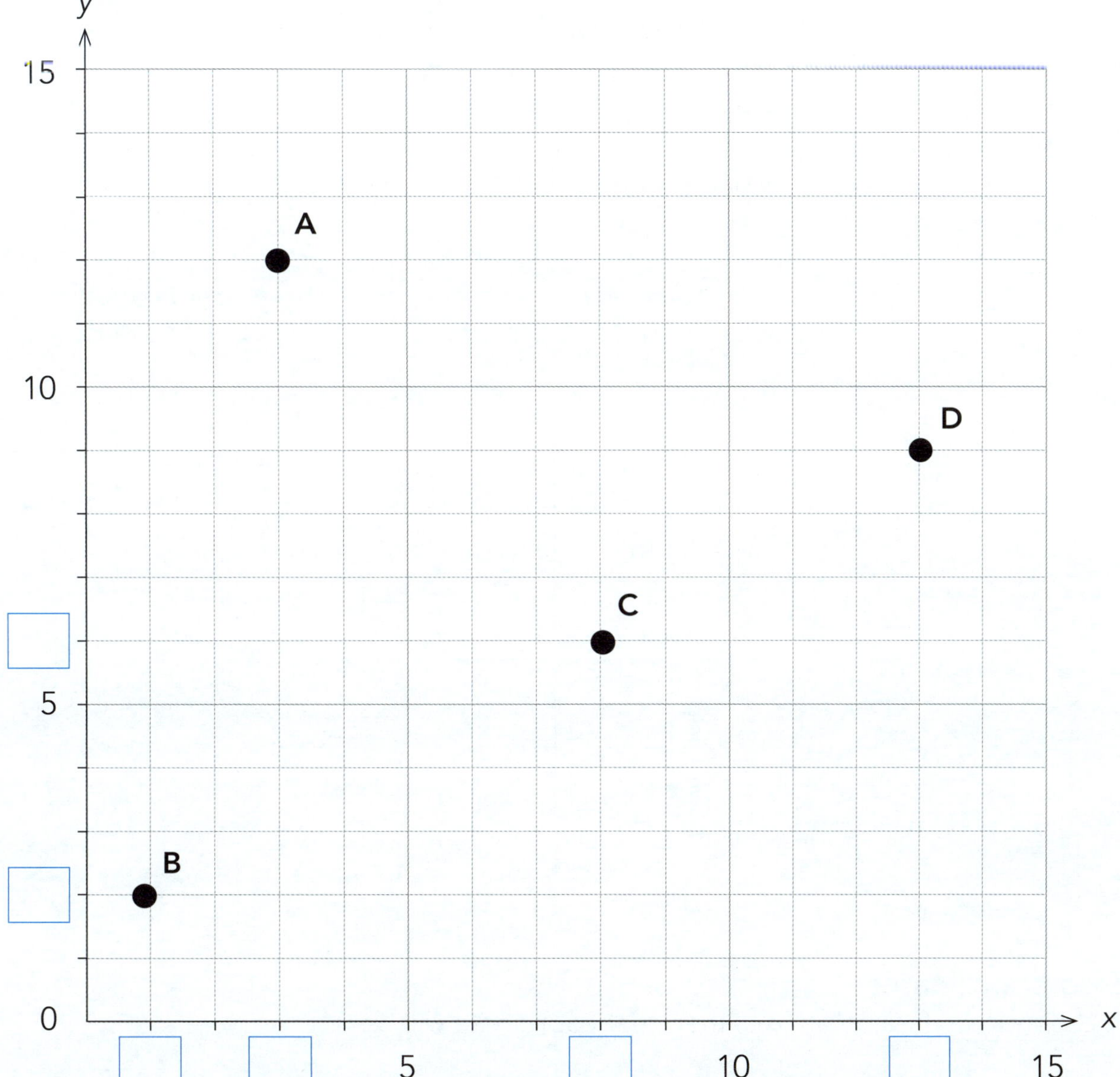

b Write the coordinates for the points A to D.

A (______, ______) B (______, ______)

C (______, ______) D (______, ______)

c Plot and label these points on the axes.

E (7,14) F (10, 4) J (3, 8) K (11, 11)

ISBN: 9780170451505

Negative coordinates

- Some axes have negative values on them as well as positive.
- If the **x** coordinate is negative, then you move to the **left**.
- If the **y** coordinate is negative, then you move **down**.

Example:

Consider the point **A** on the graph. It would be written as **(–3, –4)**.

The **x** coordinate tells us how far to move to the **left**.

The **y** coordinate tells us how far to move **down**.

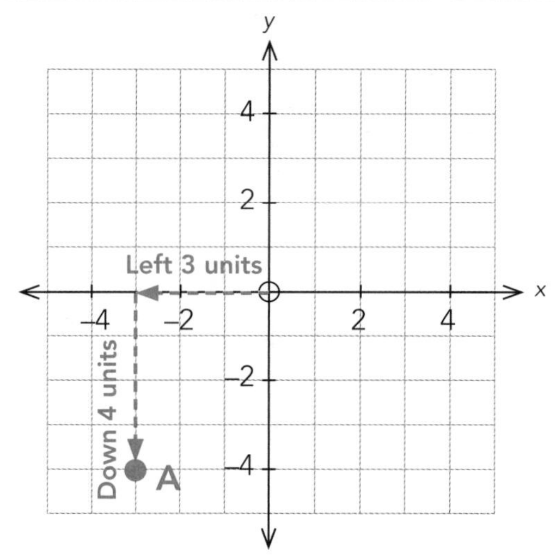

Zero coordinates

If a coordinate falls on an axis, then that coordinate will be 0.

Examples:

B (0, 3)

C (2, 0)

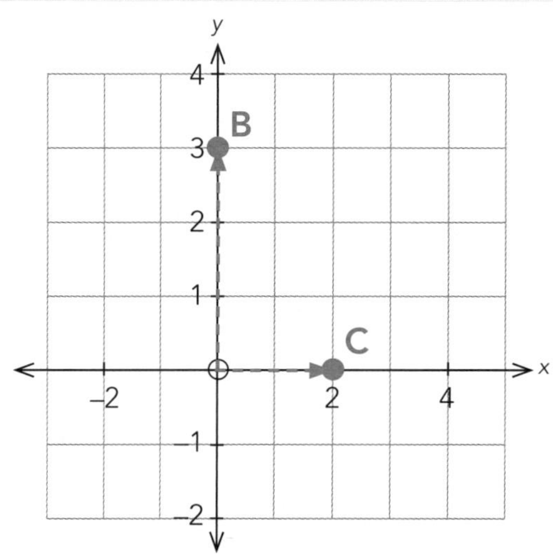

ISBN: 9780170451505

Write down the coordinates of the lettered points shown below.
The first one is done for you.

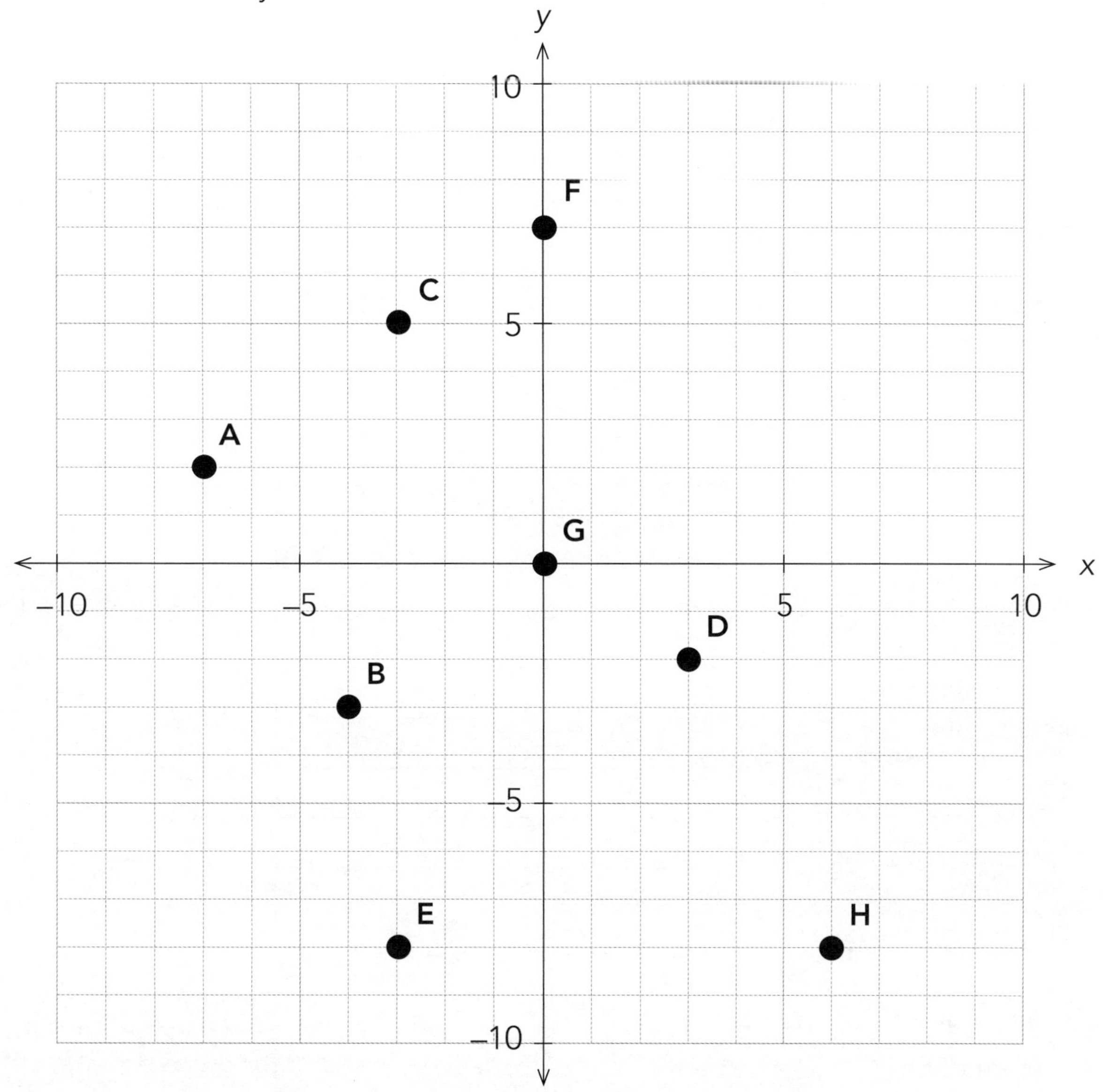

A	(–6, 2)	**B**	(______, ______)
C	(______, ______)	**D**	(______, ______)
E	(______, ______)	**F**	(______, ______)
G	(______, ______)	**H**	(______, ______)

Plot these points on the axes.

I	(–6, 6)	**J**	(10, 0)
K	(0, –3)	**L**	(–7, –5)
M	(3, –10)	**N**	(–5, 7)
O	(0, –7)	**P**	(–4, 0)

ISBN: 9780170451505

Given pattern — fill in the table, find the equation and plot the points

- You need to be able to continue a pattern and fill in a table, find the rule (equation), and use it to plot the points on a graph.
- When we plot the pattern on a graph, we use the letters **x** and **y** rather than n and T.
- While calculating term 0 is useful for finding the equation, it has no real meaning, so do not plot it.

Examples:

1 Consider the number of popsicle sticks.

Steps:

1 Put the pattern in a **table**.

For the term number, we use **x** instead of n.

We use the word **equation** rather than rule.

Term number (x)	Value of the term (y)
1	5
2	8
3	11
4	**14**
5	**17**
6	**20**

+ 3, + 3, + 3, + 3, + 3

For the value of the term, we use **y** instead of T.

2 Find the **equation**: y = 3 x term number + 2

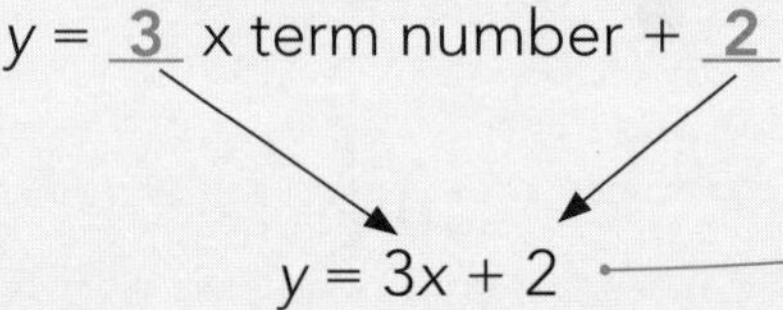

Tidy it up: $y = 3x + 2$

We use the letters **x** and **y** in equations.

3 Plot the points on a **graph**:

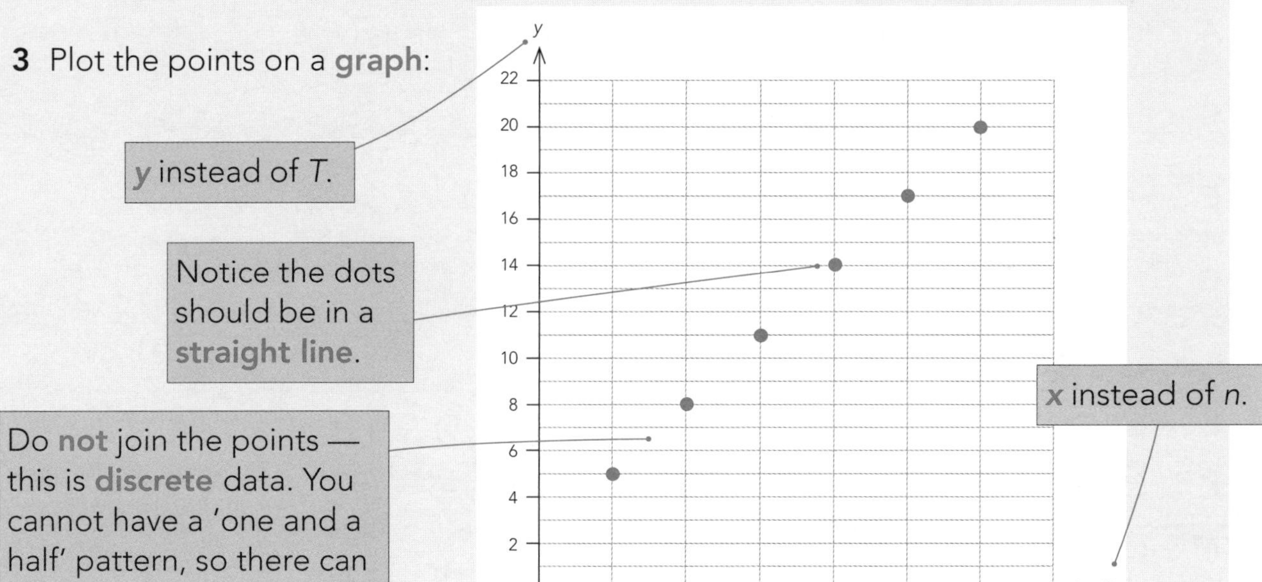

ISBN: 9780170451505

2 Consider the number of small triangles.

Steps:

1 Put the pattern in a **table**.

Term number (x)	Value of the term (y)
1	20
2	18
3	16
4	**14**
5	**12**
6	**10**

– 2
– 2
– 2
– 2
– 2

2 Find the **equation**: y = <u>–2</u> x term number + <u>22</u>

Tidy it up: $y = -2x + 22$

3 Plot the points on a **graph**:

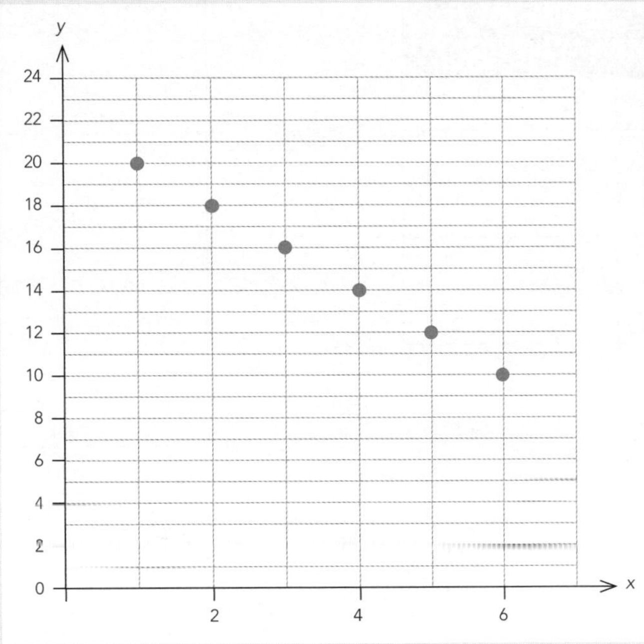

Complete each table, find the rule, and plot the points on a graph.

1 Consider the number of popsicle sticks.

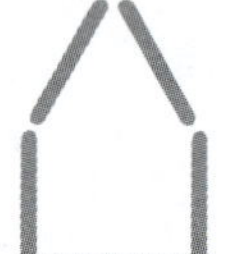

Term number (x)	Value of the term (y)
1	5
2	
3	
4	
5	

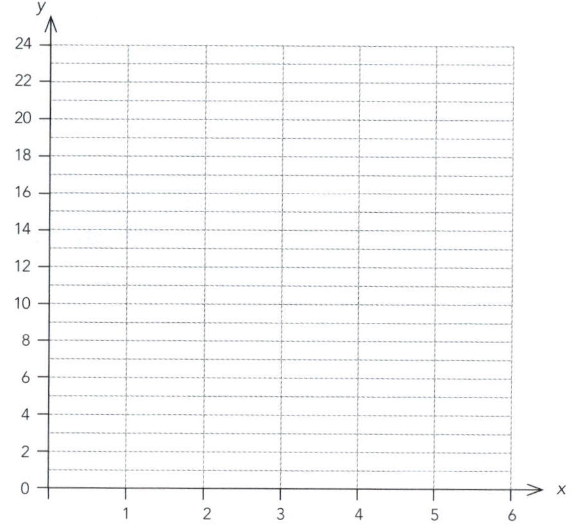

y = ______ x term number + ______

Tidy it up: y = ______x + ______

2 Consider the number of dots.

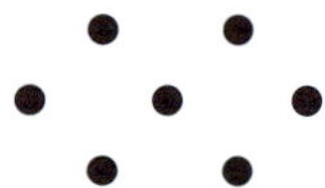
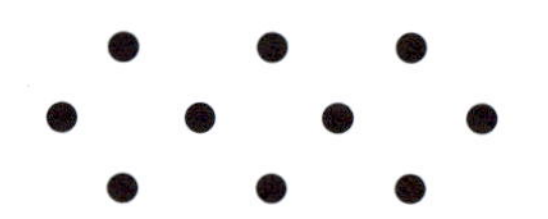
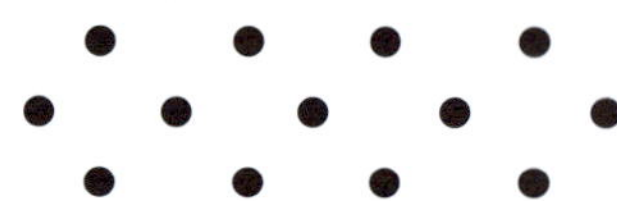

Term number (x)	Value of the term (y)
1	7
2	
3	
4	
5	

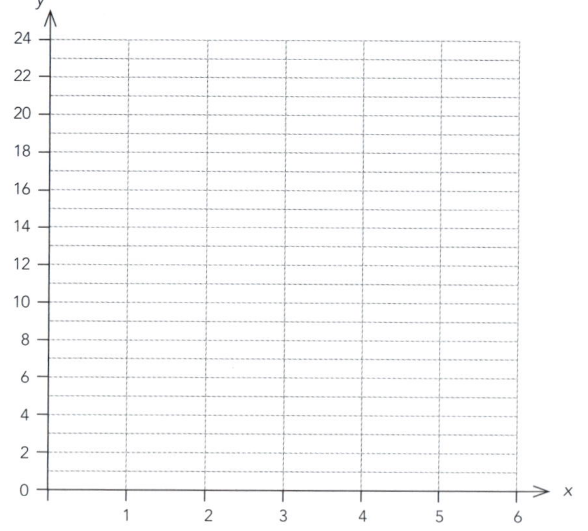

y = ______ x term number + ______

Tidy it up: y = ______x + ______

ISBN: 9780170451505

3 Consider the number of popsicle sticks.

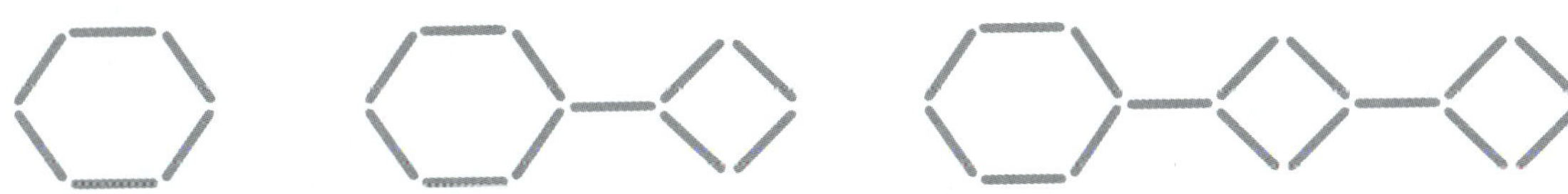

Term number (x)	Value of the term (y)
1	6
2	
3	
4	
5	**26**

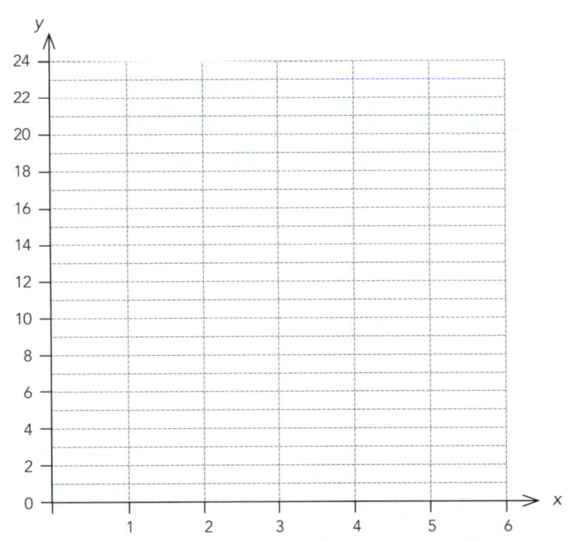

Equation: $y =$ ______ $x +$ ______

Sometimes you will get points that don't fit on the axes you are given. If this happens, just leave the point out.

4 Consider the number of crosses.

× × × ×
× × × × × ×

× × ×
× × × × ×

× ×
× × × ×

Term number (x)	Value of the term (y)
1	10
2	
3	
4	
5	

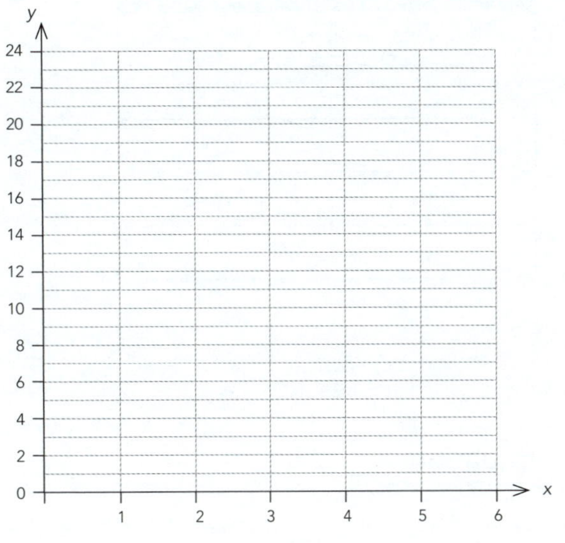

Equation: $y =$ ______ $x +$ ______

ISBN: 9780170451505

5 Consider the number of octagons.

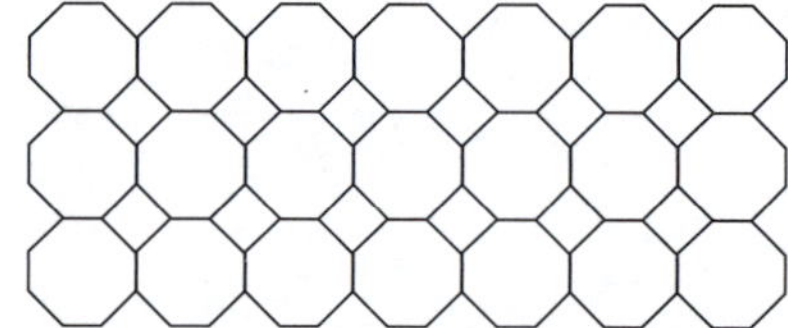
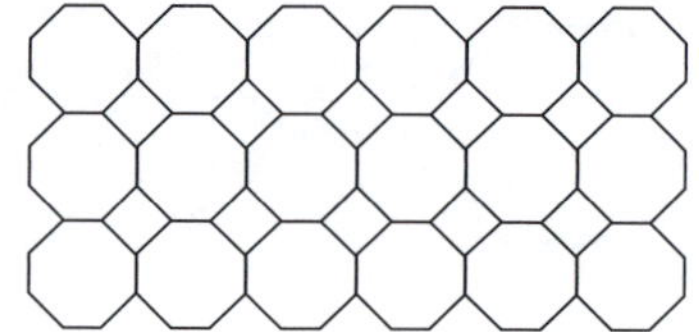
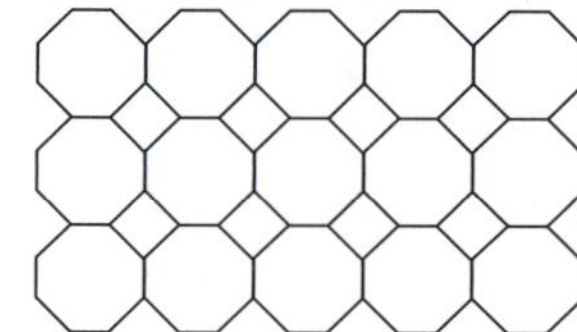

Term number (x)	Value of the term (y)
1	21
2	
3	
4	
5	

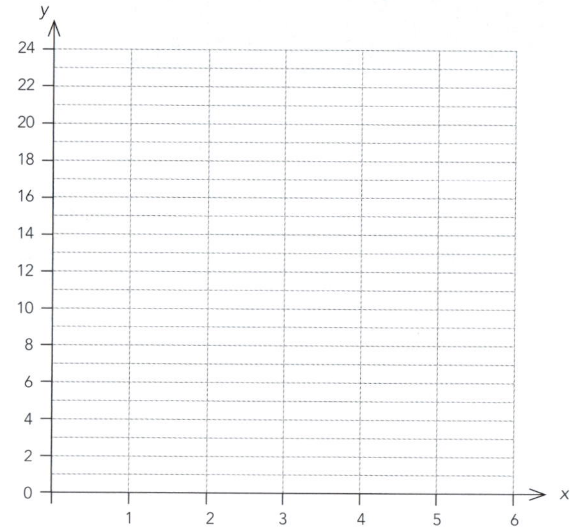

Equation: $y =$ ______ $x +$ ______

6 Ria is selling bags of plums for $3 each at the market. Her mum gives her five $1 coins so she will have change if people need it. Complete the table and graph to show how much money Ria will have after selling x bags of plums.

Bags (x)	Money (y)
1	8
2	
3	
4	
5	

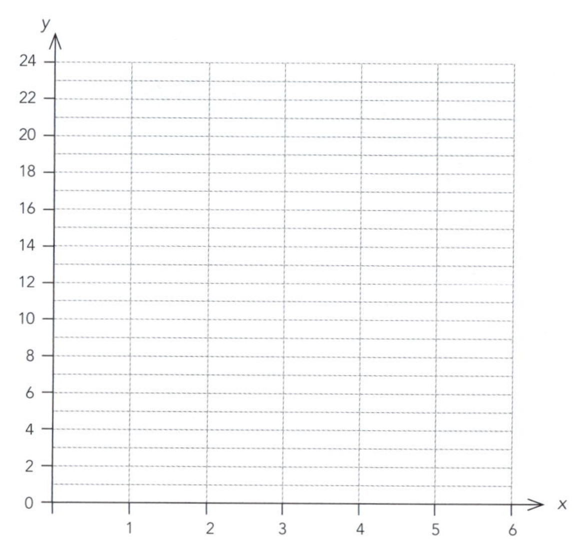

Equation: $y =$ ______ $x +$ ______

How much money will Ria have when she has sold 10 bags of plums?

____

ISBN: 9780170451505

7 Albert has a bus swipe card with \$20 credit. Each ride costs \$2. Complete the table and graph to show the credit remaining after x rides.

Ride (x)	Credit (y)
1	18
2	
3	
4	
5	

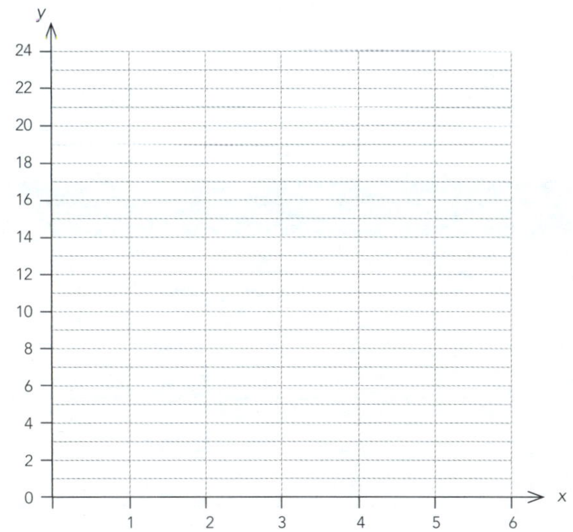

Equation: $y =$ ______ $x +$ ______

How much credit will he have after 9 rides?

____

8 Fred is building a set of steps in his garden. He needs 6 screws for each step. He has bought a packet containing 50 screws. Complete the table and graph to show the number of screws remaining after he has completed x steps.

Be careful of the scale.

Steps (x)	Screws (y)
1	
2	
3	
4	
5	

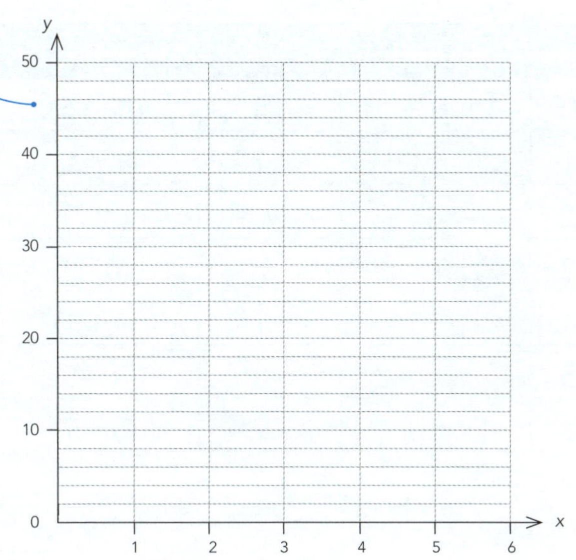

Equation: $y =$ ______ $x +$ ______

How many screws will remain after he has built 8 steps?

____

ISBN: 9780170451505

Given equation — fill in the table and plot the points

- Use a rule to plot the graph.

Examples:

1 Equation: $y = 2x - 1$

x	Calculation	y	Coordinates
0	**2 x 0 – 1**	**–1**	**(0, –1)**
1		1	(1, 1)
2		3	(2, 3)
3		5	(3, 5)
4		7	(4, 7)
5		9	(5, 9)
6	2 x **6** – 1	**11**	(6, 11)

(Each y value increases by +2.)

Check term 6.
2 x **6** – 1 = **11** ✓

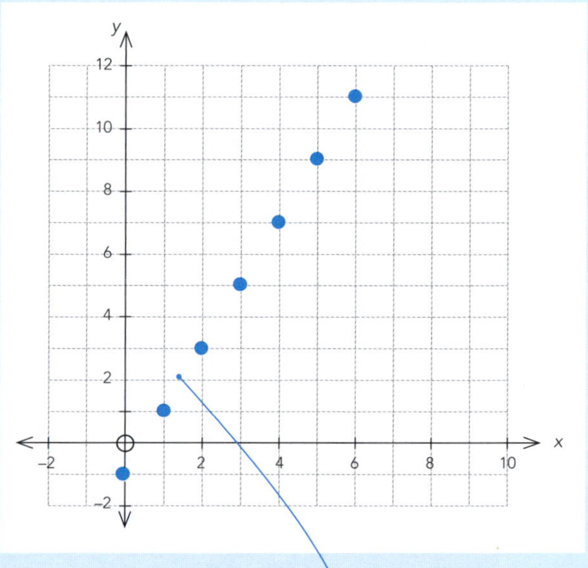

Once again, do **not** join the points — this is **discrete** data so there is no such thing as the $1\frac{1}{2}$ point.

2 Equation: $y = -4x + 15$

x	Calculation	y	Coordinates
0	**–4 x 0 + 15**	**15**	**(0, 15)**
1		11	(1, 11)
2		7	(2, 7)
3		3	(3, 3)
4		–1	(4, –1)
5		–5	(5, –5)
6	–4 x **6** + 15	**–9**	(6, –9)

(Each y value changes by –4.)

Check term 6.
–4 x **6** + 15 = **–9** ✓

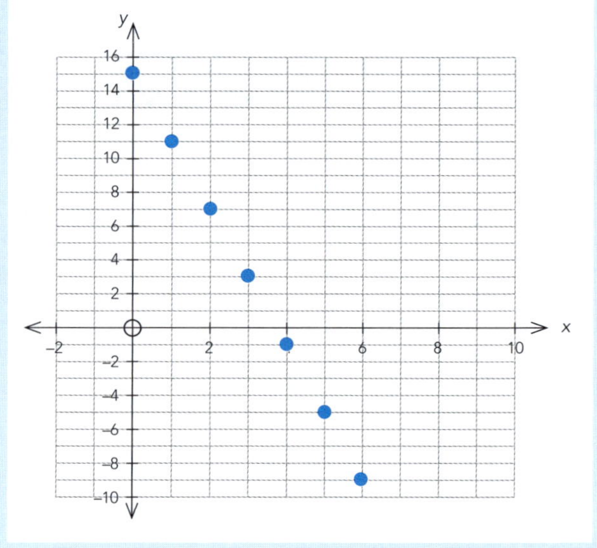

ISBN: 9780170451505

Use each equation to complete the table and then plot the points on the graph.

1 $y = 3x + 1$

x	Calculation	y	Coordinates
0	3 x 0 + 1	1	(0, 1)
1			
2			
3			
4			
5			
6			

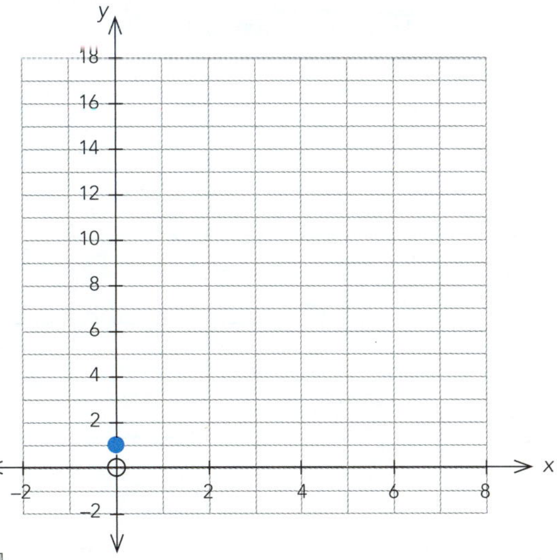

Sometimes points will not fit on the graph. As long as you have at least three points, this will not matter.

2 $y = 2x + 4$

x	Calculation	y	Coordinates
0			
1			
2			
3			
4			
5			
6			

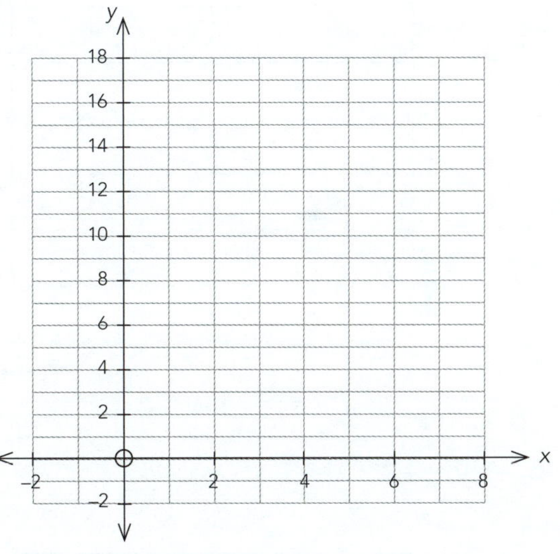

3 $y = 4x - 2$

x	y
0	
1	
2	
3	
4	
5	
6	

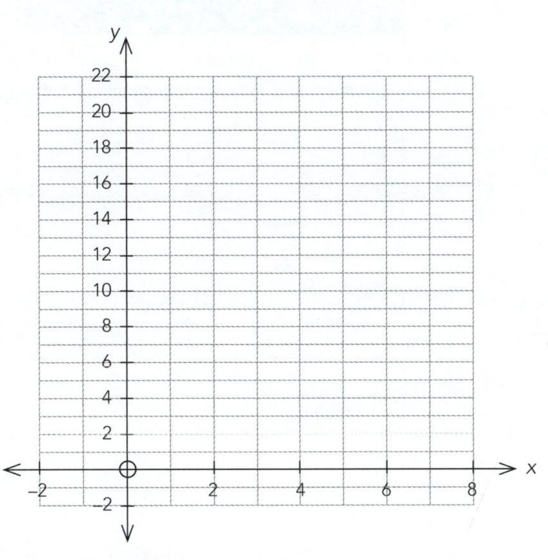

ISBN: 9780170451505

4 $y = 5x - 1$

x	y
0	
1	
2	
3	
4	
5	
6	

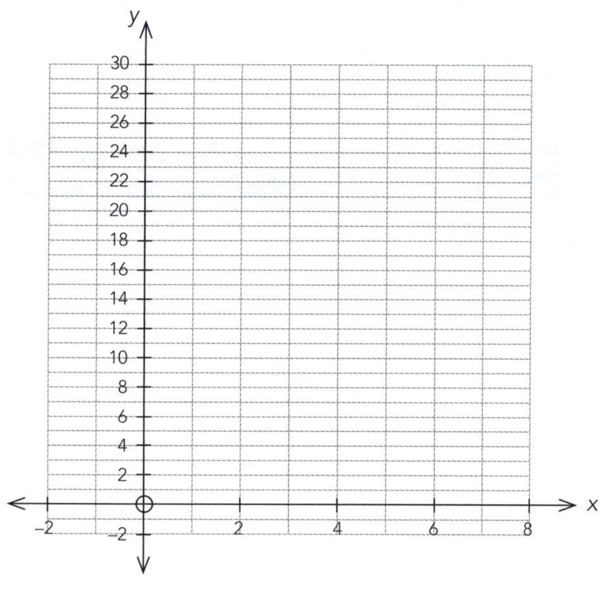

5 $y = -2x + 8$

x	y
0	
1	
2	
3	
4	
5	
6	

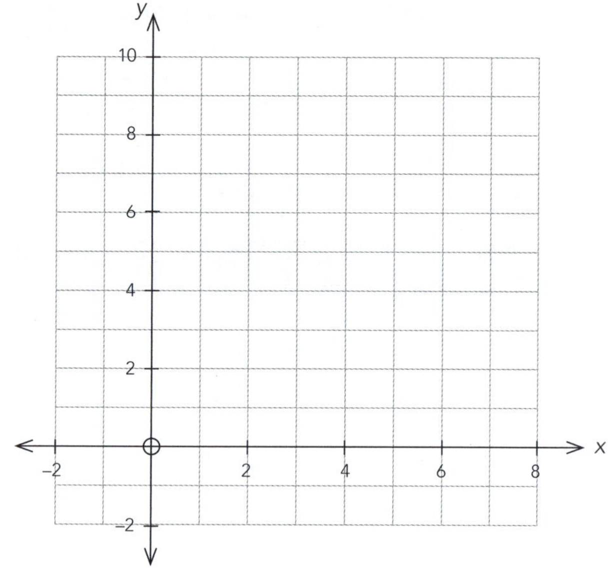

6 $y = -x + 7$

x	y
0	
1	
2	
3	
4	
5	
6	

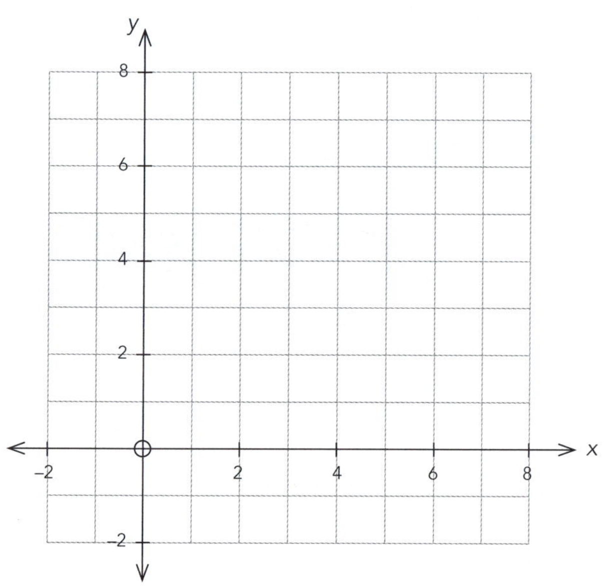

ISBN: 9780170451505

7 $y = -x - 3$

x	y
0	
1	
2	
3	
4	
5	
6	

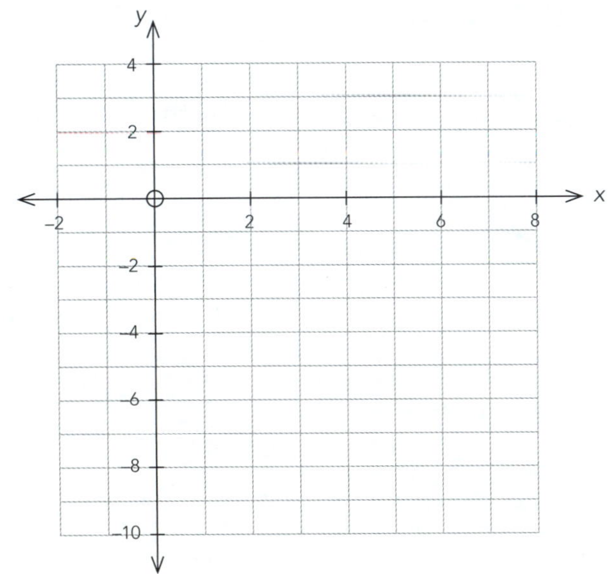

8 $y = 2x$

x	y
0	
1	
2	
3	
4	
5	
6	

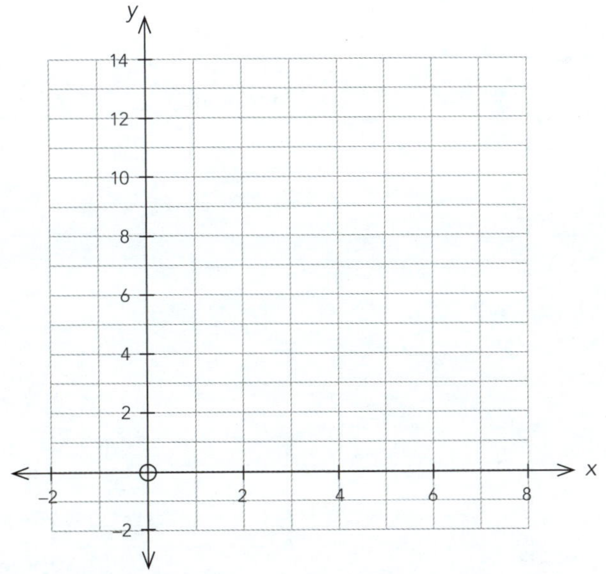

9 $y = -x$

x	y
0	
1	
2	
3	
4	
5	
6	

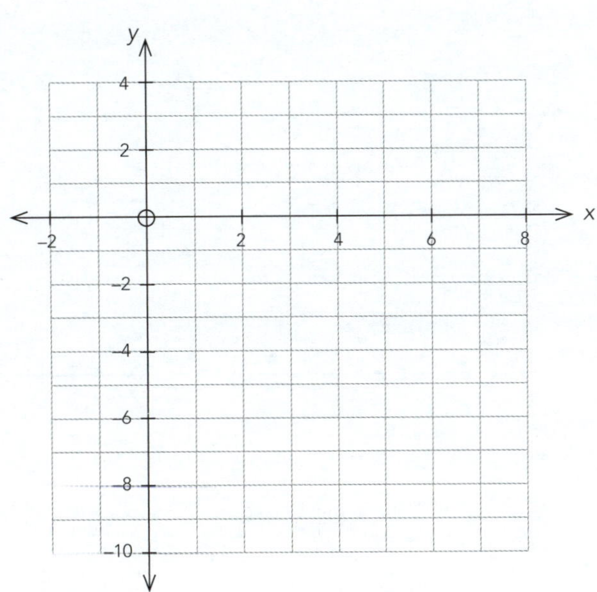

ISBN: 9780170451505

Given graph — fill in the table and write the equation

- You need to be able to use the information on a graph to fill in a table and write an equation.

Example:

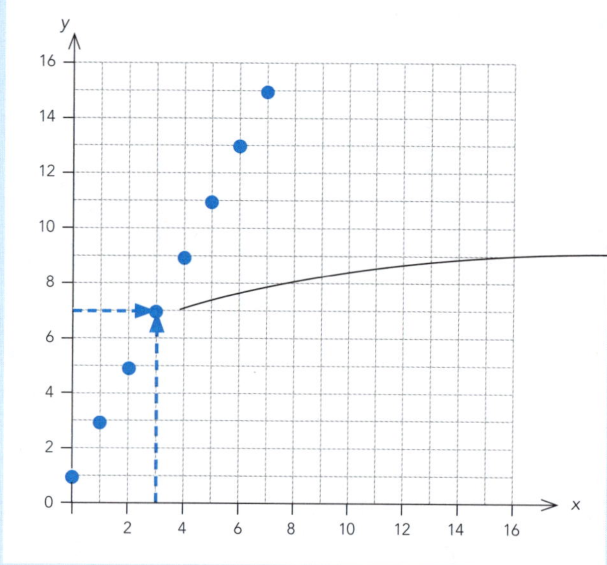

x	y
0	1
1	3
2	5
3	7
4	9
5	11
6	13

y = 2 x term number + 1

Equation: $y = 2x + 1$

1

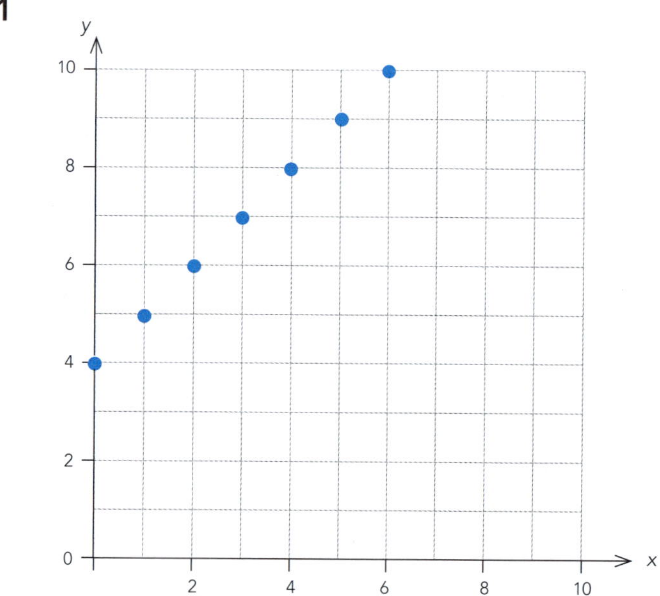

x	y
0	
1	
2	
3	
4	
5	
6	

Equation: $y =$ ______ $x +$ ______

 ISBN: 9780170451505

2

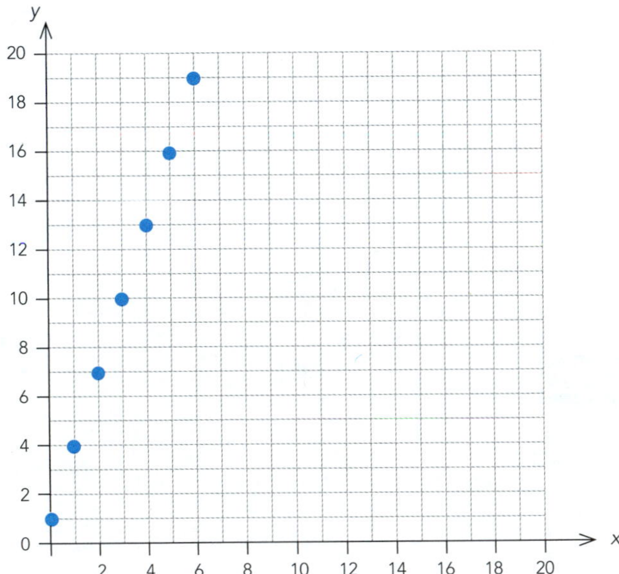

x	y
0	
1	
2	
3	
4	
5	
6	

Equation: $y =$ ______ $x +$ ______

3

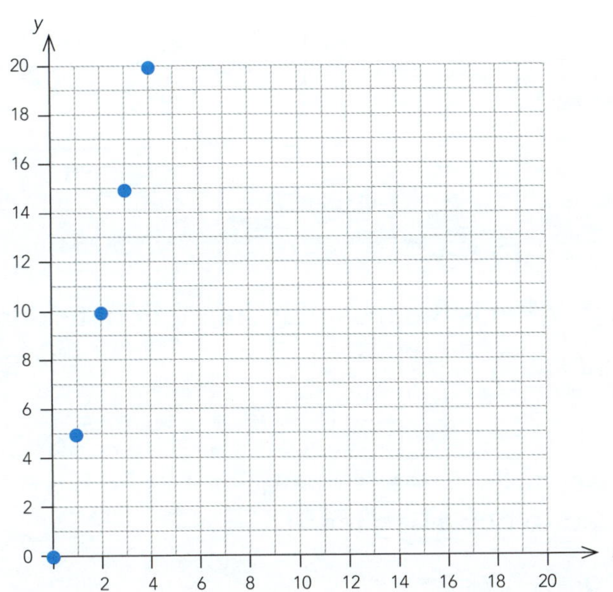

x	y
0	
1	
2	
3	
4	

Equation: $y =$ ______ $x +$ ______

4

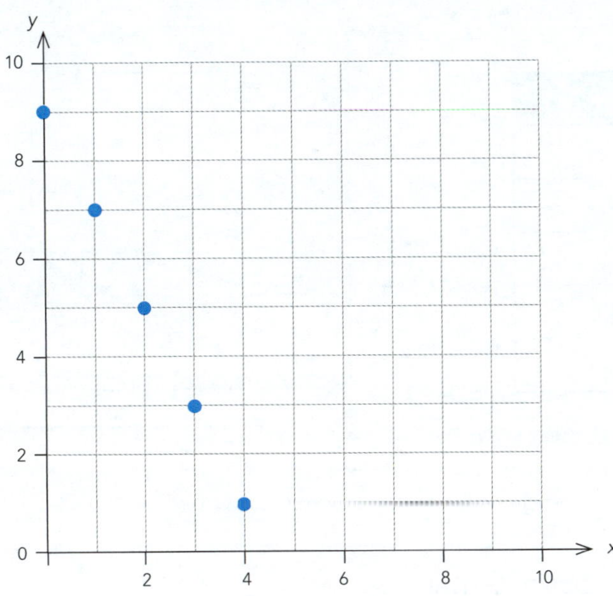

x	y
0	
1	
2	
3	
4	

Equation: $y =$ ______ $x +$ ______

ISBN: 9780170451505

Applications

Use the skills that you have learnt so far to answer these questions.

1 Mallory offered to buy concert tickets for herself and her friends. There is a booking fee of \$2 and each ticket costs \$5.

a Complete the table and plot the points to show the cost of buying t tickets.

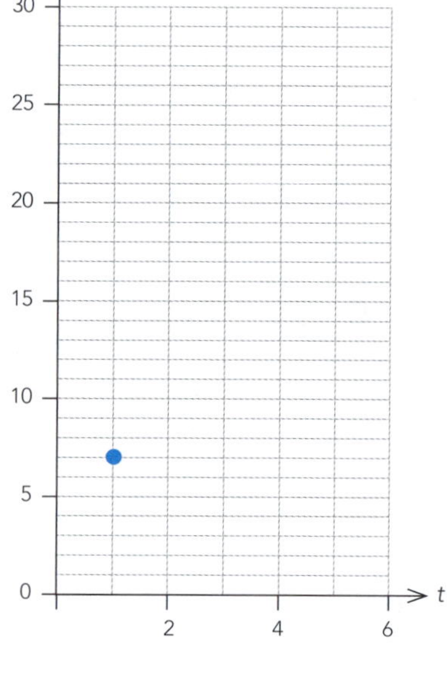

Number of tickets (t)	Cost (c)
1	
2	
3	
4	
5	

b Write an equation to show the cost of t tickets (when she buys at least one ticket).

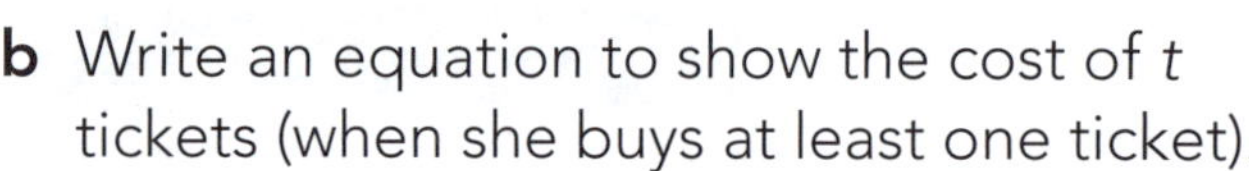

$c =$ ______ $t +$ ______

c Use the equation to calculate the cost of 7 tickets. ____________

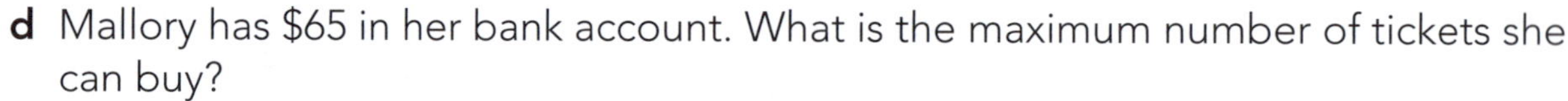

d Mallory has \$65 in her bank account. What is the maximum number of tickets she can buy?

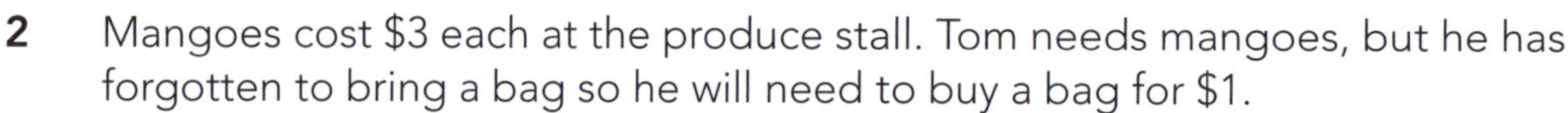

____

2 Mangoes cost \$3 each at the produce stall. Tom needs mangoes, but he has forgotten to bring a bag so he will need to buy a bag for \$1.

a Complete the table and plot the points to show the cost of buying a bag and m mangoes.

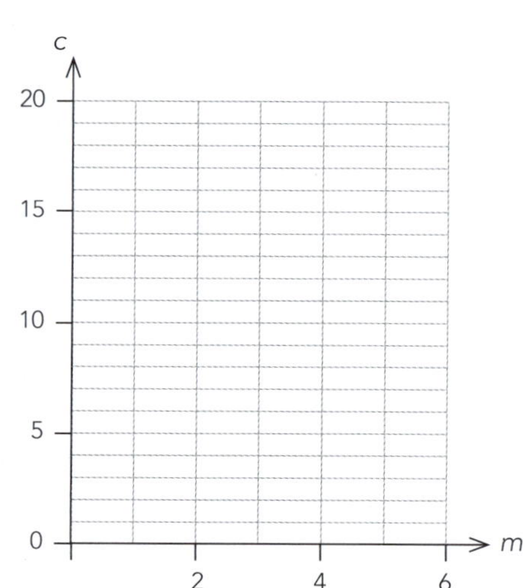

Number of mangoes (m)	Cost (c)
1	
2	
3	
4	
5	

b Write an equation to show the cost of m mangoes and a bag.

$c =$ ______ $m +$ ______

c Use the equation to calculate the cost of 9 mangoes and a bag.

____

 ISBN: 9780170451505

3 Grandma has a packet containing 30 buttons. She is knitting jackets for each of her baby grandchildren. Each jacket needs 4 buttons.

a Complete the table and plot the points to show the number of buttons remaining after she has completed *j* jackets.

Number of jackets (j)	Number of buttons remaining (b)
1	
2	
3	
4	
5	

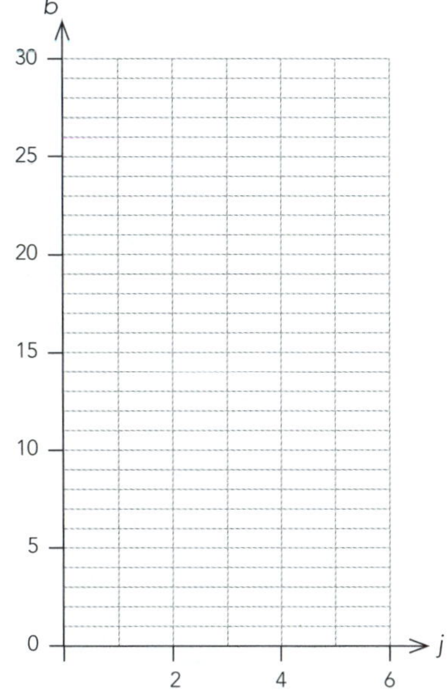

b Write an equation to show the number of buttons remaining after she has completed *j* jackets.

$b =$ ______ $j +$ ______

c Use the equation to calculate the number of buttons remaining after Grandma has completed 7 jackets.

____

d How many extra buttons would she need in order to complete 9 jackets? ______

4 Edwina has a doggy daycare business. She charges $10 a day plus $4 for each hour or part of an hour that she has the dog.

a Complete the table and plot the points to show the total cost for caring for a dog for *h* hours.

Time (hours) (h)	Cost ($) ($c$)
1	
2	
3	
4	
5	

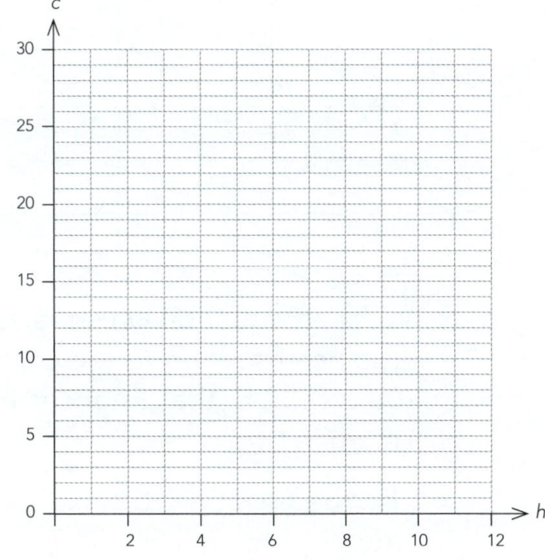

b Write an equation that relates the number of hours (or parts of hours) that a dog is with Edwina to the amount it will cost.

____

c If Geraldine leaves her dog with Edwina for 8 hours, how much will that cost?

____

ISBN: 9780170451505

5 Tom's Taxis charges a flat $10 fee plus $2 for each complete kilometre.

a The blue dots on the graph show the cost of a ride in Chloe's Cabs. Use them to complete this table.

Distance (km) (d)	Cost (c)
2	
4	
6	
8	
10	
12	

b Write an equation which relates the distance travelled to the cost of the ride.

$c =$ ____________________

c Explain how the cost of a ride in Chloe's Cabs is calculated.

____

____

d Complete the table and add the points to the graph to show the total cost (c) of a ride of d kilometres in one of Tom's Taxis.

Distance (km) (d)	Cost (c)
2	
4	
6	
8	
10	
12	
14	

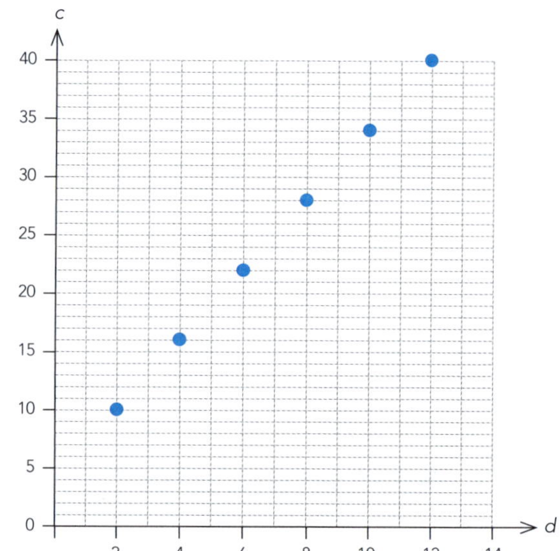

e Write an equation which relates the distance travelled to the cost of the ride. (Be careful — the distances in the table go up in twos!)

$c =$ ____________________

f Use the equation to calculate how much a ride of 21 km will cost in Tom's Taxis.

____

____

g Write down the distance for which the cost of rides in Tom's Taxis and Chloe's Cabs are the same. How much do they cost?

____

 ISBN: 9780170451505

Drawing straight lines

- Up until now we have drawn lines of dots on graphs. This is because we have been dealing with **discrete data**.
- **Discrete data** is data that is **counted**, e.g. number of tickets, number of avocados, etc.
- **Continuous data** is **measured**, e.g. length, area, volume, mass, time, etc.
- With continuous data, you can have fractions, so the gaps between discrete values can be joined by a line.
- There are several ways of plotting straight lines. Drawing up a table is a method that works for any graph, including curves.

Example:
Plot the line given by the equation $y = 2x + 4$.

Equations are the same as rules. Often they are written using x and y, but sometimes we use different letters.

Step 1: Fill in the values for x and y.

Term (x)	2x + 4	Answer (y)	Point
0	2 x **0** + 4	4	(0, 4)
1		6	(1, 6)
2		8	(2, 8)
3		10	(3, 10)
4		12	(4, 12)
5	2 x **5** + 4	14	(5, 14)

(+2 between each successive answer)

Always plot **at least three points**. If they don't form a straight line, check your calculations.

Step 2: Plot the points.

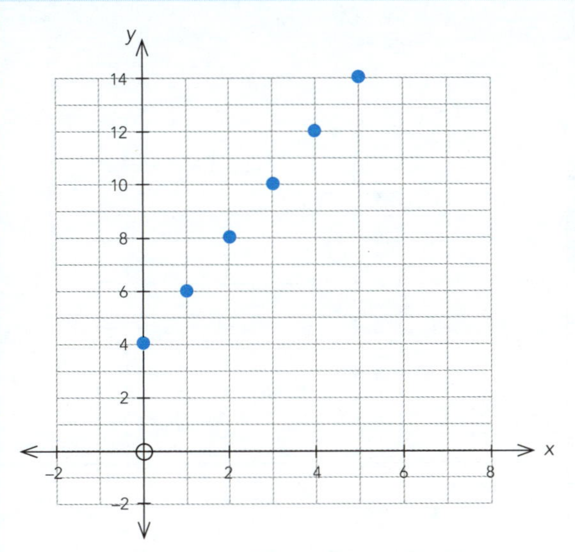

Step 3: Join the points with a **ruled** line.

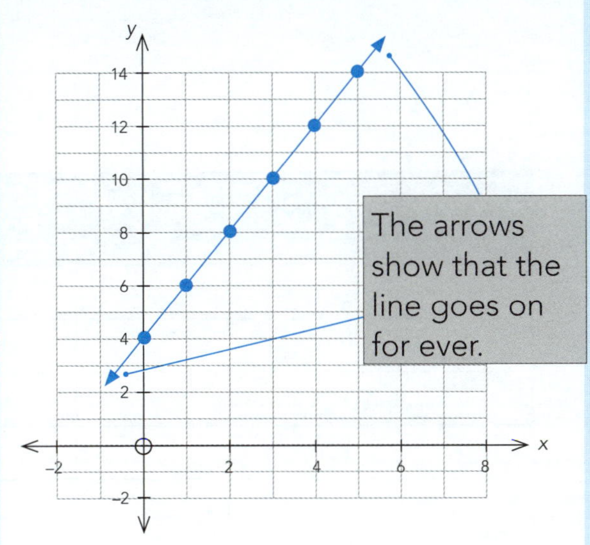

ISBN: 9780170451505

Complete each table, plot the points, and join them with a ruled line.

1 $y = 3x + 2$

Term (x)	3x + 2	Answer (y)	Point
0	3 x 0 + 2	2	
1		5 (+3)	(1, 5)
2			
3			
4			

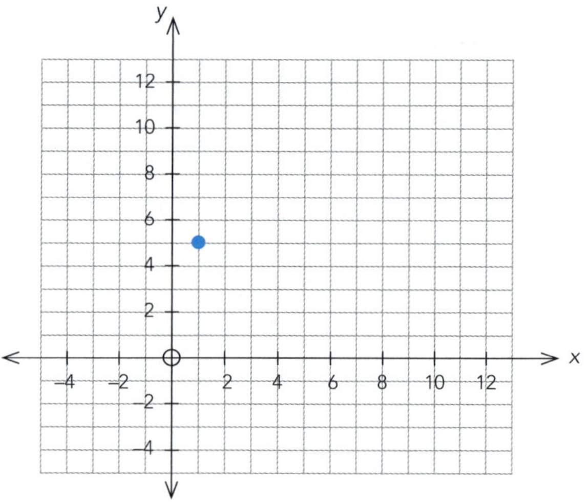

2 $y = 2x + 1$

Term (x)	2x + 1	Answer (y)	Point
0			
1			
2			
3			
4			

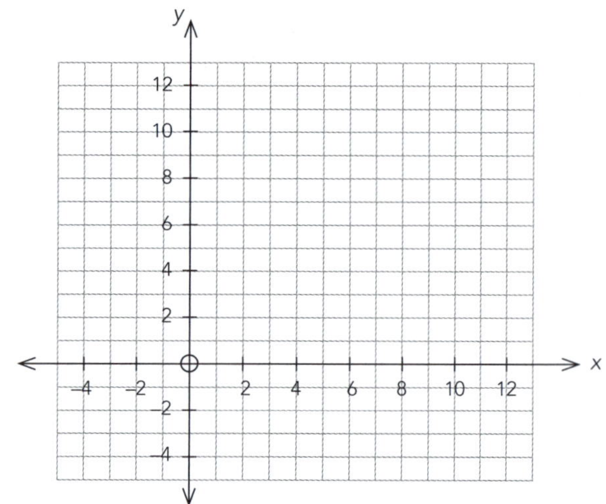

3 $y = x + 4$

Term (x)	x + 4	Answer (y)	Point
0			
1			
2			
3			
4			

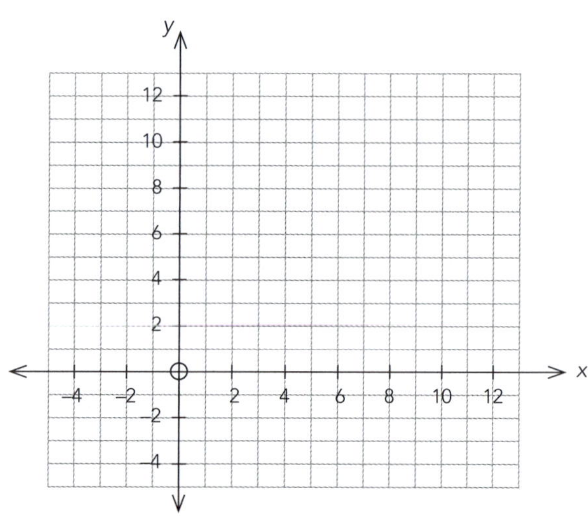

ISBN: 9780170451505

4 $y = -2x + 6$

Term (x)	Answer (y)	Point
0		
1		
2		
3		
4		

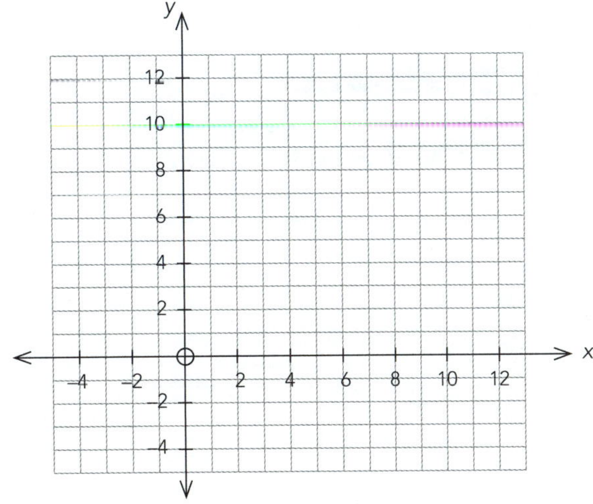

5 $y = -x + 8$

Term (x)	Answer (y)	Point
0		
1		
2		
3		
4		

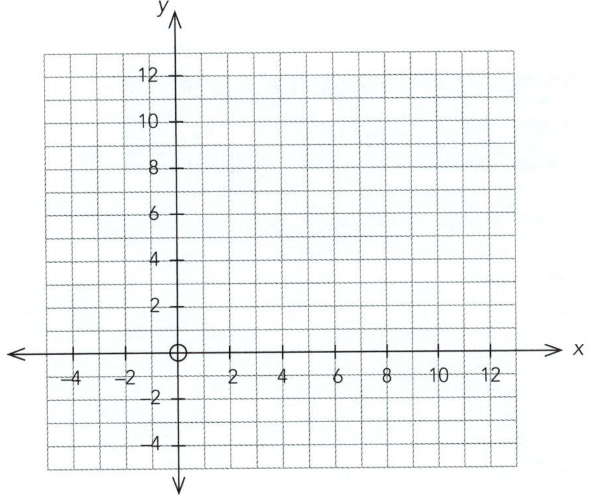

6 $y = 4x$

Term (x)	Answer (y)	Point
0		
1		
2		
3		
4		

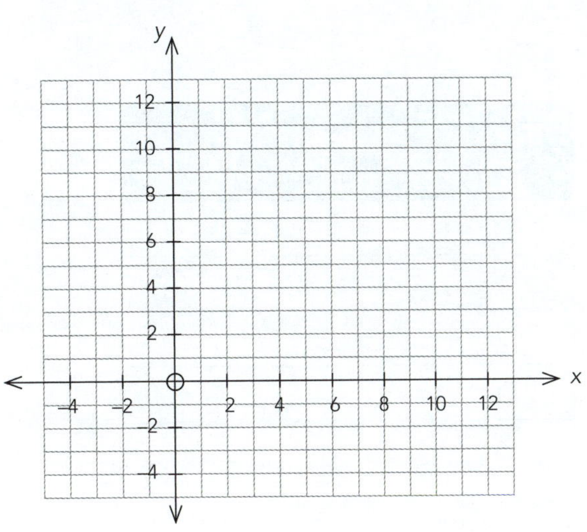

ISBN: 9780170451505

7 $y = 2x - 3$

Term (x)	Answer (y)	Point
0		
1		
2		
3		
4		

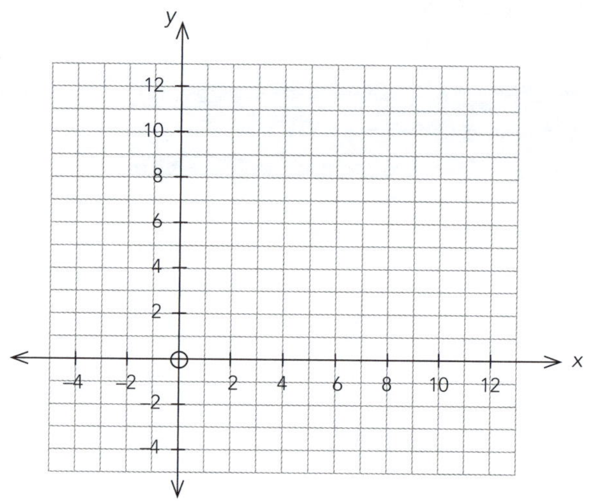

8 $y = x - 4$

Term (x)	Answer (y)	Point
0		
1		
2		
3		
4		

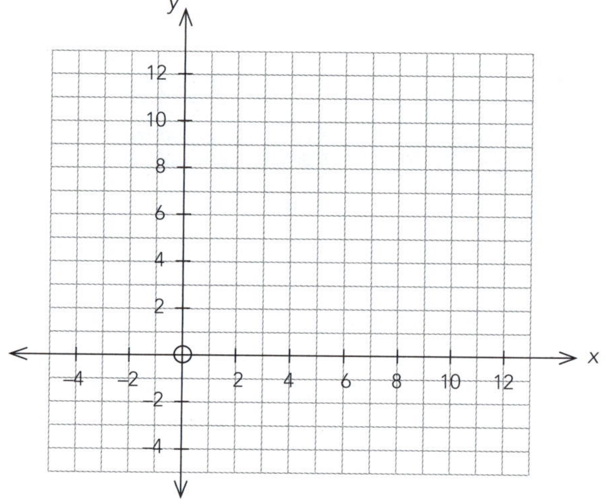

9 $y = -x + 5$

Term (x)	Answer (y)	Point
0		
1		
2		
3		
4		

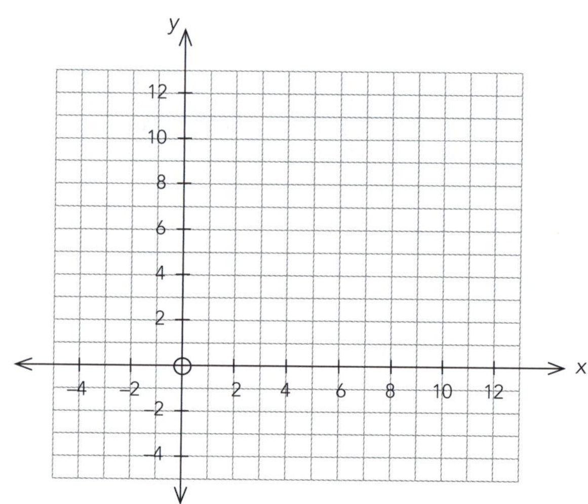

ISBN: 9780170451505

Given graph — write down the coordinates and find the equation

- Sometimes you will be given a line on a graph and you will need to find the equation.
- Usually you will be given just the line without the points.

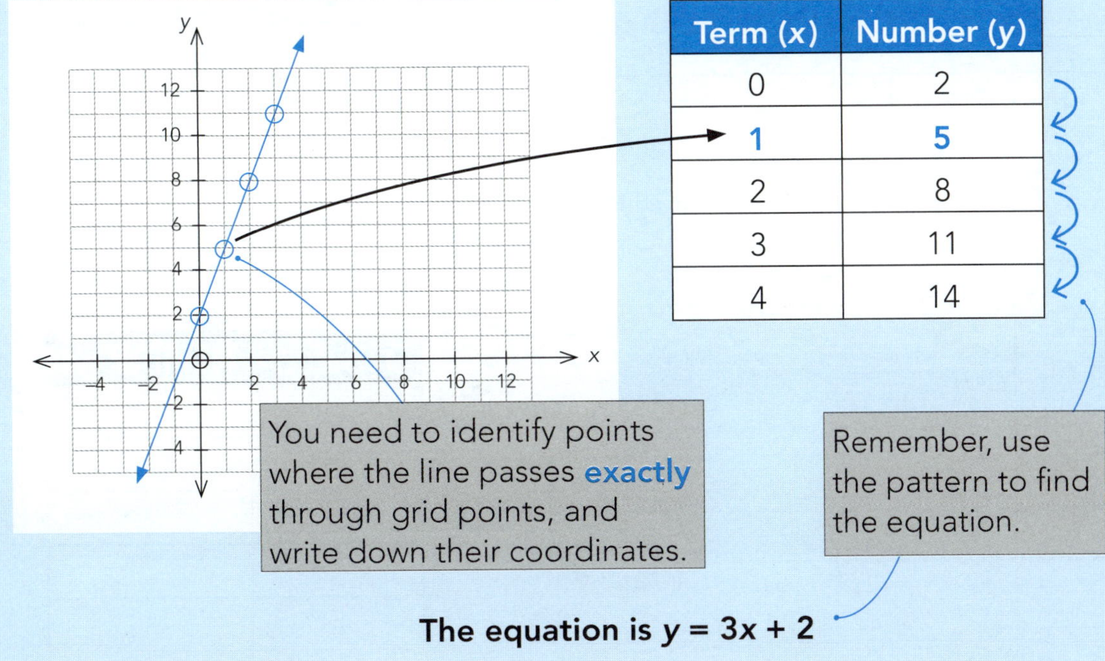

Term (x)	Number (y)
0	2
1	**5**
2	8
3	11
4	14

The equation is $y = 3x + 2$

Complete each table and find the equation for the line.

1

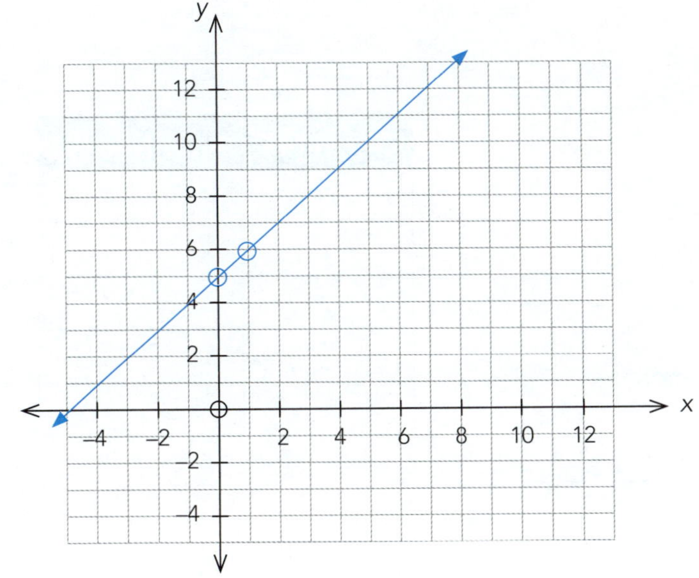

Term (x)	Number (y)
0	**5**
1	**6**
2	
3	
4	

The equation is $y =$ ______$x +$ ______

2

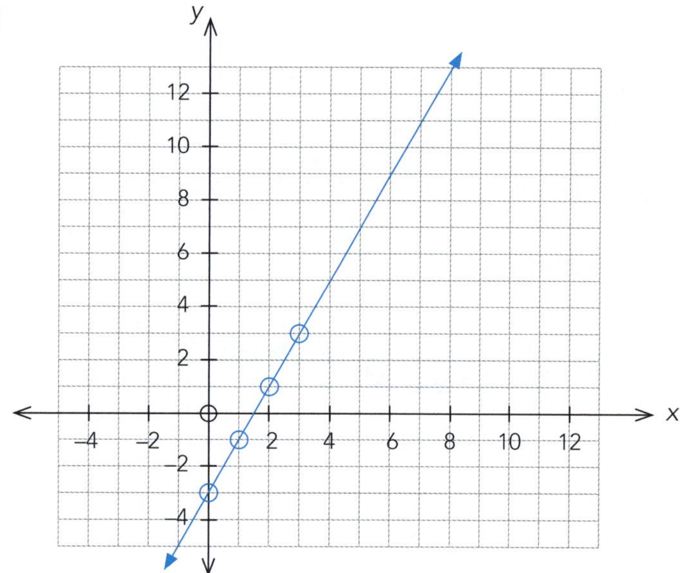

Term (x)	Number (y)
0	–3
1	
2	
3	
4	

The equation is y = ______________

3

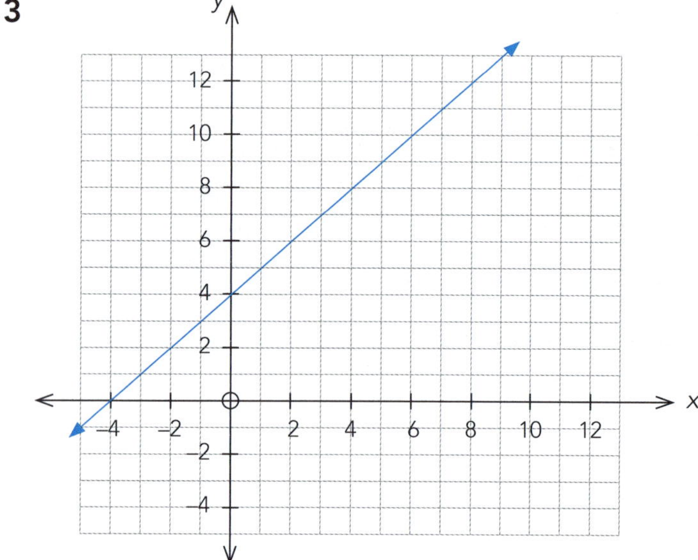

Term (x)	Number (y)
0	
1	
2	
3	
4	

The equation is y = ______________

4

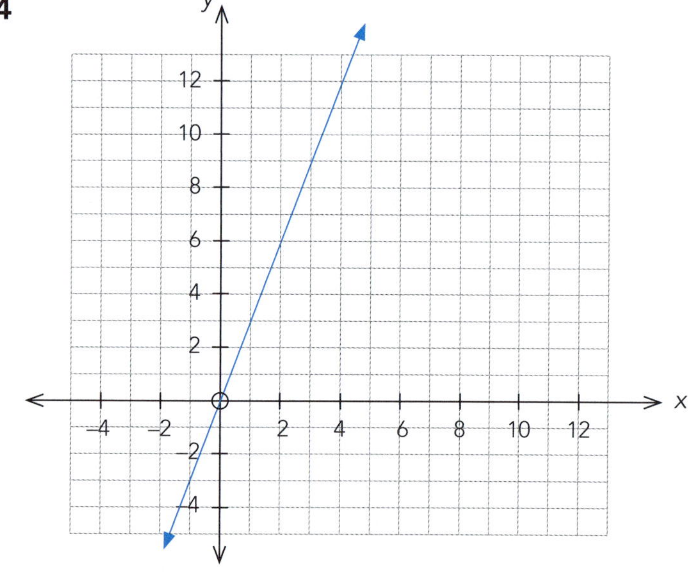

Term (x)	Number (y)
0	
1	
2	
3	
4	

The equation is y = ______________

ISBN: 9780170451505

5

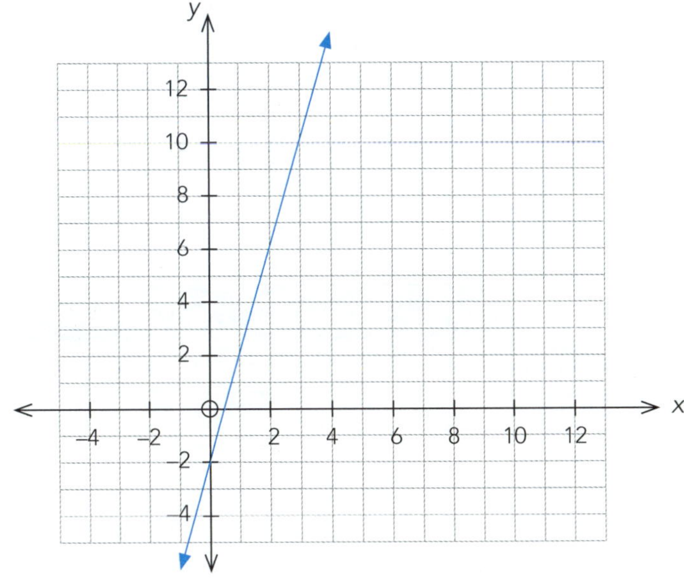

Term (x)	Number (y)
0	
1	
2	
3	
4	

The equation is $y =$ ____________

6

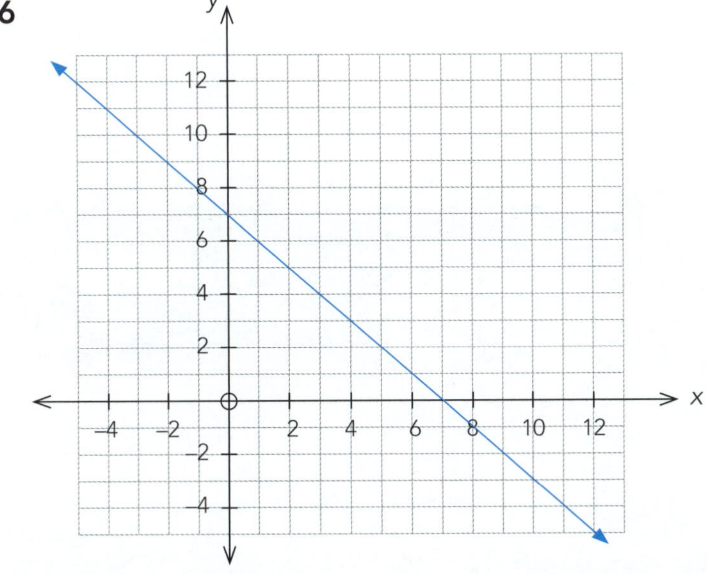

Term (x)	Number (y)
0	
1	
2	
3	
4	

The equation is $y =$ ____________

7

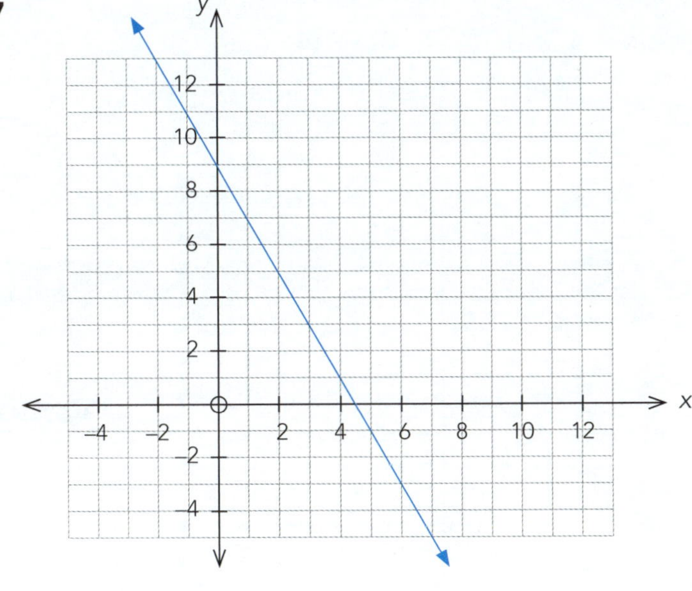

Term (x)	Number (y)
0	
1	
2	
3	
4	

The equation is $y =$ ____________

ISBN: 9780170451505

Horizontal and vertical lines

- Horizontal and vertical lines have only **one letter** in their equation.
- Be very careful drawing these — it's easy to get them the wrong way round.
- You can draw up a table, or just write a list of **at least three** points.

Examples:

1 $y = 3$

Step 1: Complete the table for $y = 3$.

It doesn't matter what the x value is, y will **always be 3**.

x	y	Coordinates
0	3	(0, 3)
1	3	(1, 3)
2	3	(2, 3)
3	3	(3, 3)

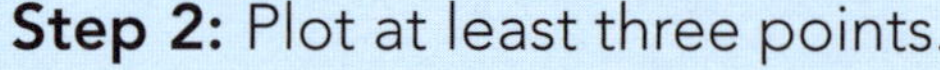

Step 2: Plot at least three points.

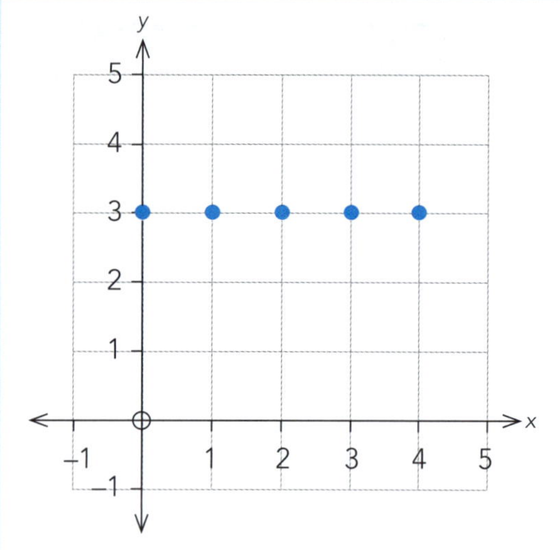

Step 3: Join the points.

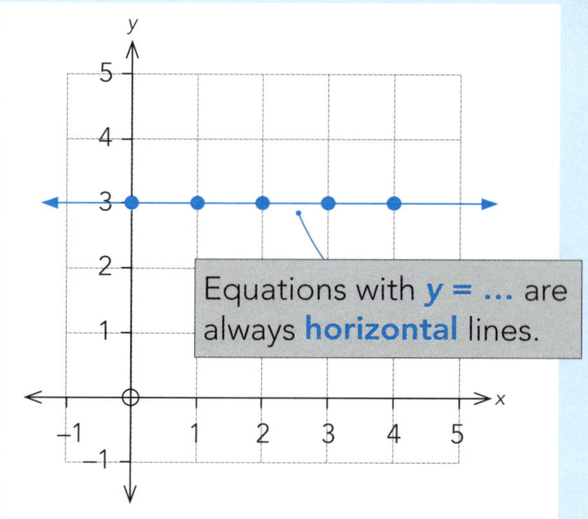

2 $x = 4$

It doesn't matter what the y value is, x will **always be 4**.

Step 1: List at least three points where $x = 4$, e.g. (4, 0), (4, 1), (4, 2), (4, 3).

Step 2: Plot the points.

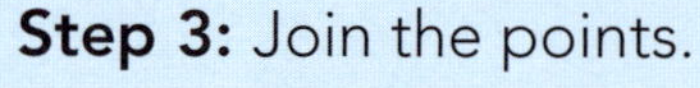

Step 3: Join the points.

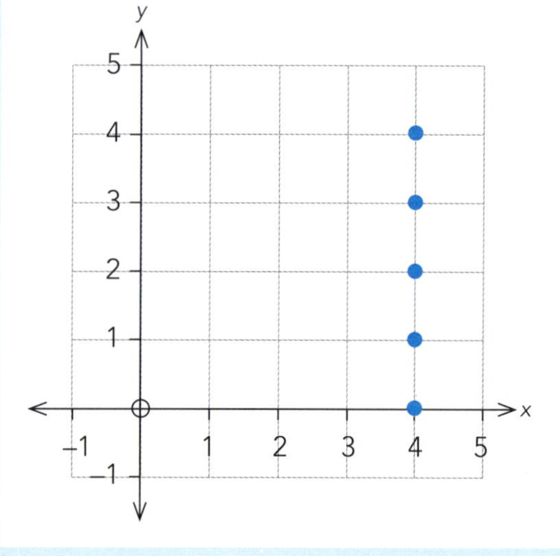

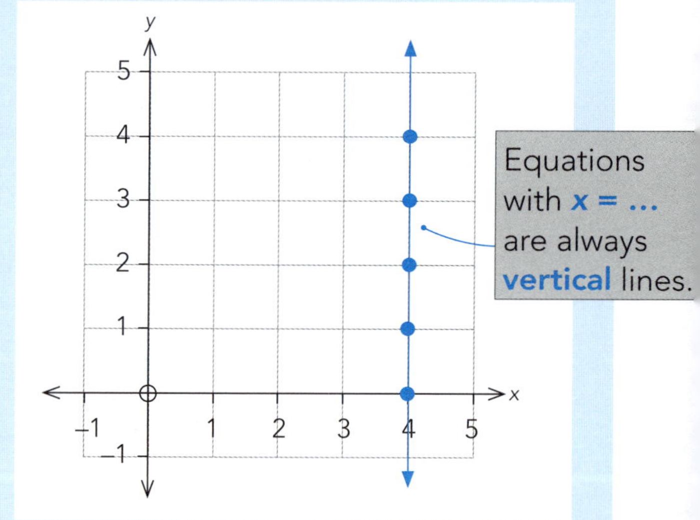

 ISBN: 9780170451505

Highlight/circle the correct equation for each graph.

1

$x = 3$

$y = -3$

$x = -3$

$y = 3$

2

$x = -1$

$y = -1$

$x = 1$

$y = 1$

3

$x = -6$

$y = 6$

$x = 6$

$y = -6$

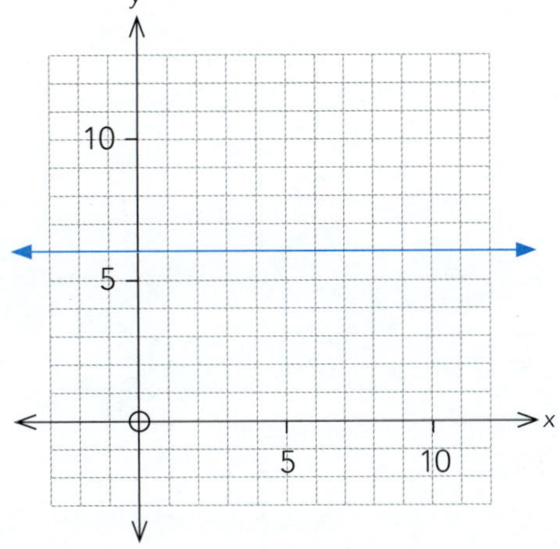

4

$x = -9$

$y = 9$

$x = 9$

$y = -9$

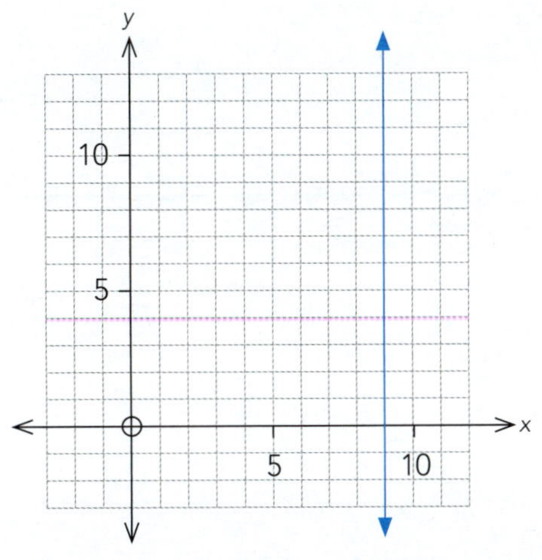

ISBN: 9780170451505

Draw these lines on the graphs.

5 $y = 2$

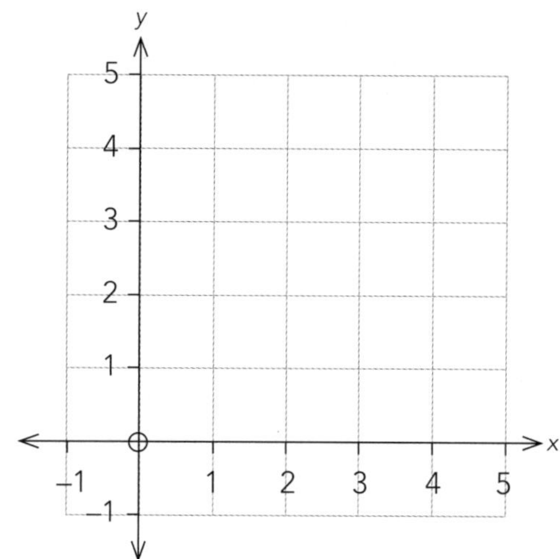

6 $x = 1$

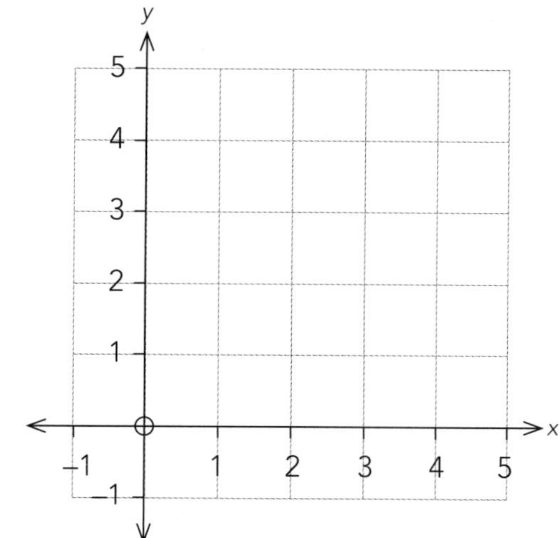

7 $x = -3$

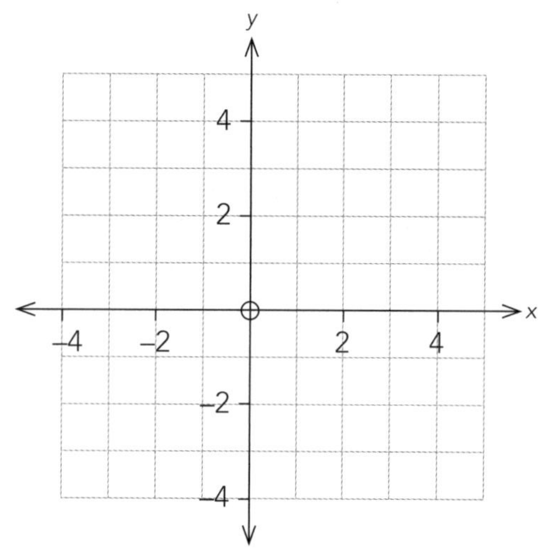

8 $y = -1$

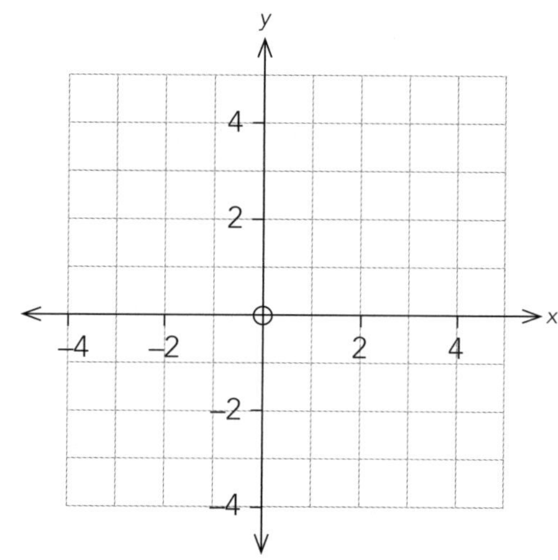

9 $y = 0$

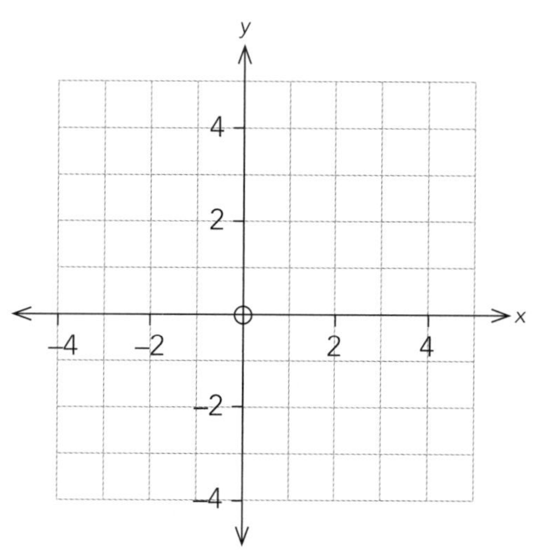

10 $x = 0$

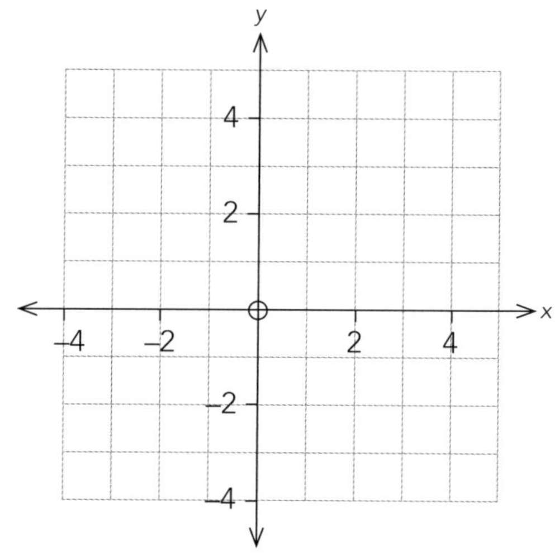

 ISBN: 9780170451505

Applications

- Graphs of straight lines have many applications.

Example: Tama's cat has had kittens. He weighs the kittens every week to check that they are gaining mass. The vet has given him a graph to show him the minimum mass for kittens as they grow older.

a Use the graph to help you complete the table to show the minimum mass for kittens as they grow.

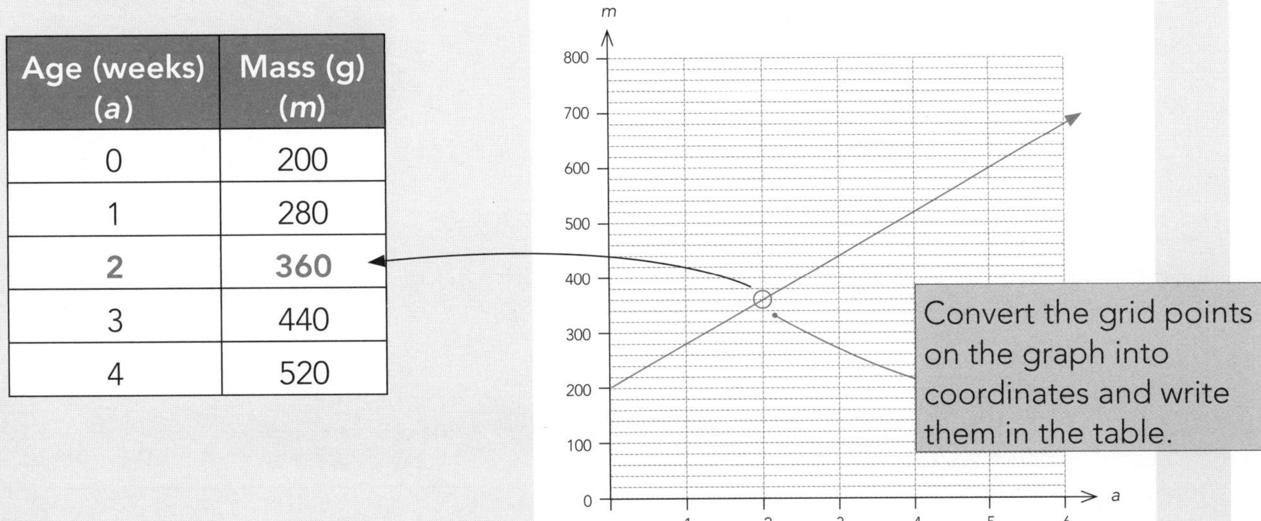

Age (weeks) (a)	Mass (g) (m)
0	200
1	280
2	**360**
3	440
4	520

b Write an equation that relates the mass of the kitten to its age.

Age (weeks) (a)	Mass (g) (m)
0	**200**
1	280
2	360
3	440
4	520

– 80
+ 80
+ 80
+ 80

$\boldsymbol{m = 80a + 200}$

c Use the equation to calculate the minimum mass of a kitten that is **6** weeks old.

$m = 80 \times \mathbf{6} + 200$
$= 680$ g

Substitute **6** for **a** (number of weeks) in the equation.

d According to the vet's graph, how much mass should a kitten gain each week?

Mass gained should be 80 g

You can read this from the graph or the table.

ISBN: 9780170451505

Use the skills that you have learnt so far to answer these questions.

1 A one-metre-deep tank with vertical sides has some water in the bottom and it is being filled at a constant rate.

a Use the graph to help you complete the table to show the depth of the water (d) after t minutes.

Time (min) (t)	Depth (cm) (d)
0	
1	
2	
3	
4	
5	

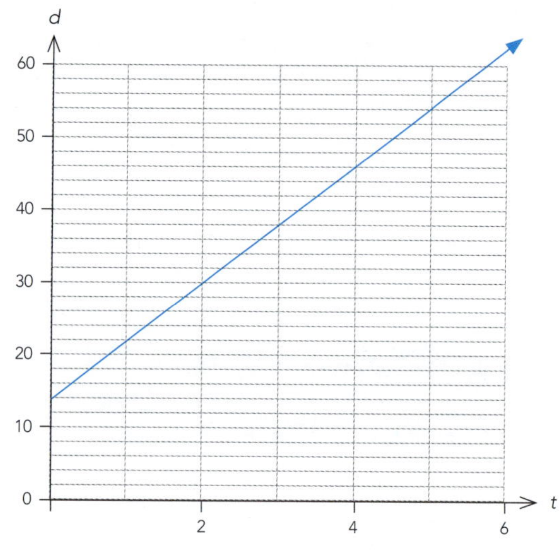

b Write an equation that relates the depth of the water to the time spent filling the tank. $d =$ ______

c Use the equation to calculate how much water was in the tank to start with. ______

d How fast is the depth of water increasing? ______

2 Jack is saving to go on a music trip. The graph shows how much he has saved so far after w weeks.

a Use the graph to help you complete the table to show how much he has saved so far.

Week (w)	Saved (\$) (s)
0	
1	
2	
3	
4	
5	

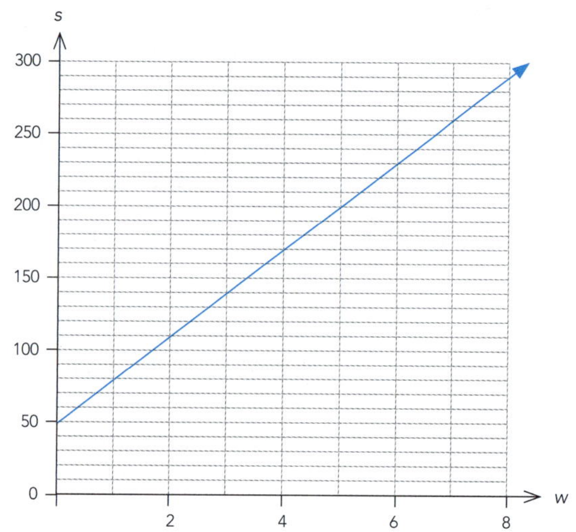

b Write an equation that relates the amount he has saved to the number of weeks since he started saving. $s =$ ______

c Use the equations to work out much he will have saved after 9 weeks. ______

 ISBN: 9780170451505

3 Kiri has borrowed money from her sister to buy some headphones. The graph shows how much she still owes her sister (o) after w weeks.

a Use the graph to help you complete the table to show how much she still owes her sister.

Week (w)	Owes ($) ($o$)
0	
1	
2	
3	
4	
5	

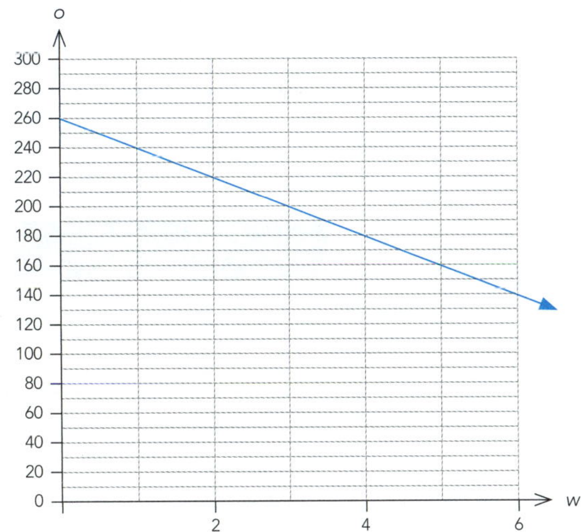

b Write an equation that relates the amount Kiri owes to the number of weeks since she bought the headphones. $o =$ ______________

c Use the equation to calculate how much Kiri owes after 10 weeks. ______________

d How much does she pay off each week? ______________

4 Bernie lives 1800 m from school and he walks home from school at an average speed of 50 m per minute.

a Complete the table and plot the points to show how far he is from school after t minutes.

Time (min) (t)	Distance from school (d)
0	
1	
2	
3	
4	
5	

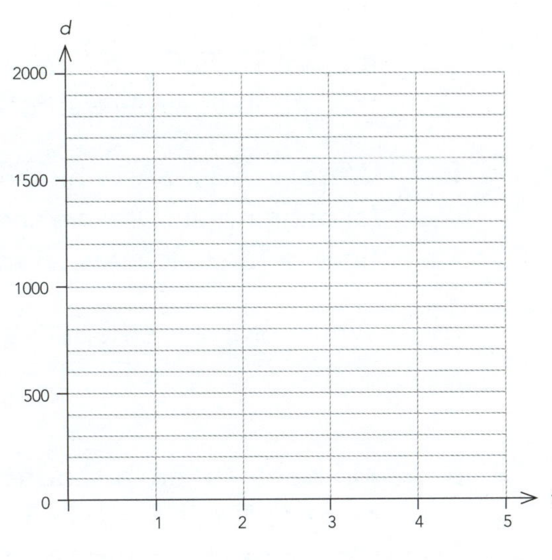

b Write an equation that relates the distance Bernie is from school to the time he has been walking. $d =$ ______________

c Use the equation to work out how far he is from school when he has walked for 12 minutes. ______________

ISBN: 9780170451505

5 Hiring an electric scooter costs a fixed basic amount plus a rate per minute. The total cost is calculated in **cents**. The blue line shows the total cost (c) of hiring a Lemon scooter for t minutes.

a Use the graph to help you complete the table to show how much it costs (c) to hire a Lemon scooter for these periods of time (t).

Time (min) (t)	Cost (c) (c)
2	
4	
6	
8	
10	
12	

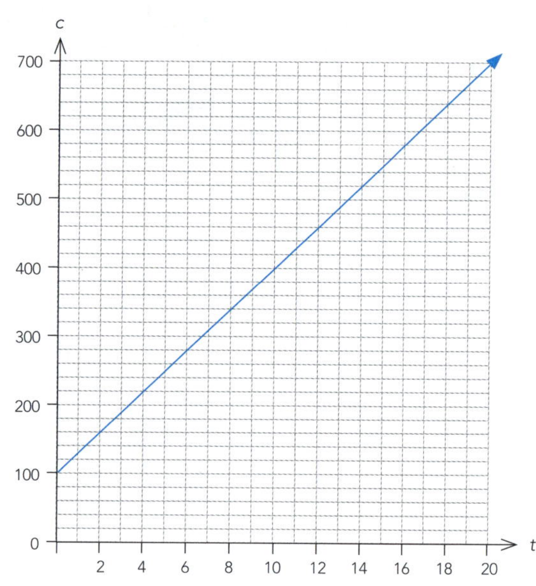

b Write an equation which relates the cost (c) of hiring a Lemon scooter to the time (t) for which it is hired. $c =$ ____________

c Use the equation to calculate how much it would cost to hire a scooter for half an hour. ____________

d How much does it cost per minute to hire a Lemon scooter? ____________

e It costs $2 (200c) to hire an Orange scooter, plus 20c per minute. Draw a line on the graph to show the total cost of hiring an Orange scooter.

f Write an equation which relates the cost (c) of hiring an Orange scooter to the time (t) for which it is hired. $c =$ ____________

g Which graph is steeper? Why? ____________

h Kiri wants to hire a scooter for 18 minutes. Draw lines on the graph to show how she could find the cost of hiring each type of scooter.

Lemon scooter: ____________

Orange scooter: ____________

i The lines cross at (10, 400). Explain what this means. ____________

 ISBN: 9780170451505

Linear or not?

- **Linear** patterns are patterns that increase or decrease by the **same amount** each time.
- A patterns that increases by **different amounts** each time is called a **non-linear** pattern.

Examples:

1 Consider the number of dots.

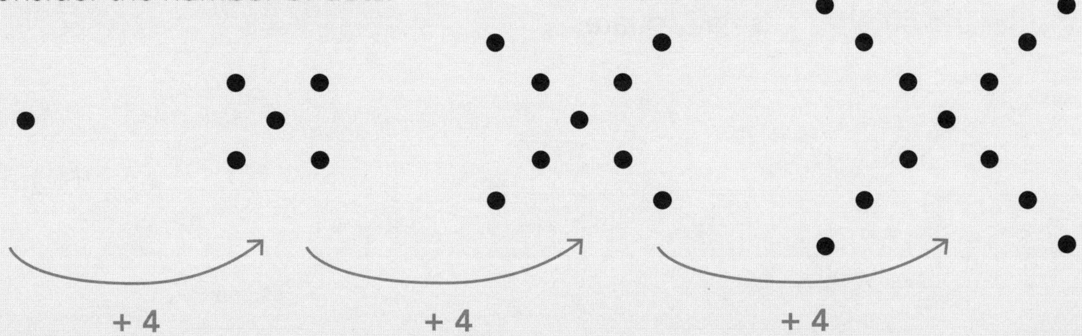

This pattern increases by the **same number** (4) each time, therefore it **is** linear.

2 Consider the number of popsicle sticks.

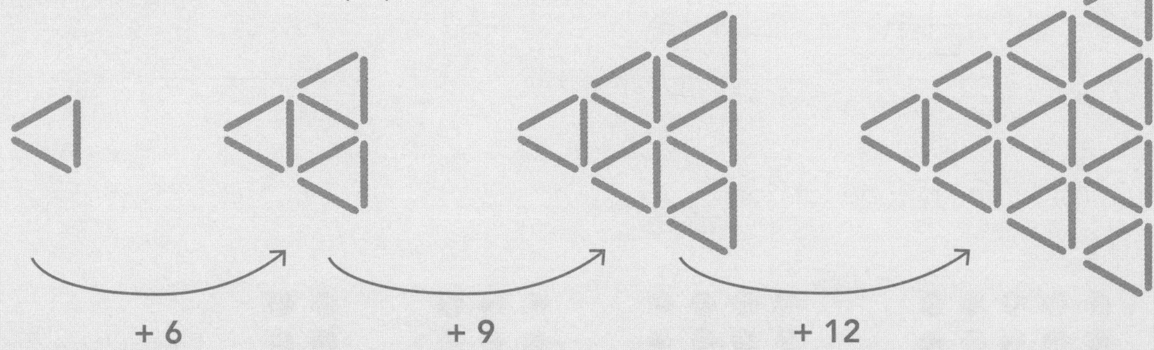

This pattern increases by **different numbers** of popsicle sticks each time, therefore it is **not** linear.

3 20, 16, 12, 8, 4

−4 −4 −4 −4

This pattern decreases by the **same number** (4) each time, therefore it **is** linear.

4 2, 4, 9, 16, 25

+2 +5 +7 +9

This pattern increases by **different numbers** each time, therefore it is **not** linear.

Identify if these patterns are linear or non-linear.

1 Consider the number of crosses.

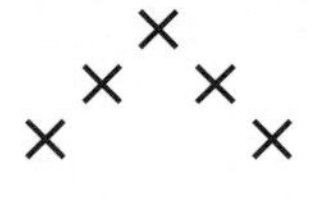 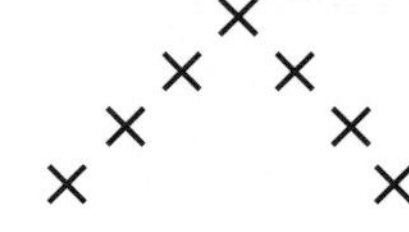 

Linear/Non-linear

2 Consider the number of small squares.

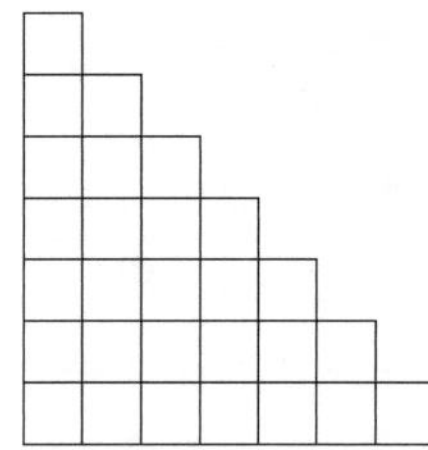 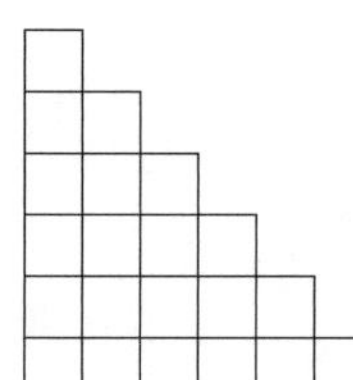 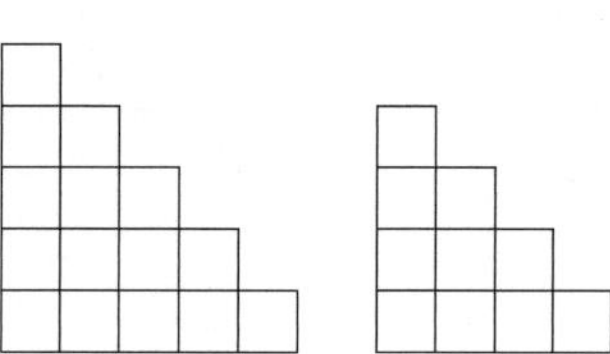

Linear/Non-linear

3 Consider the number of small squares.

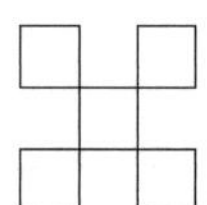 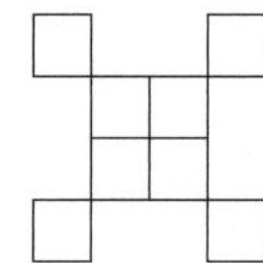 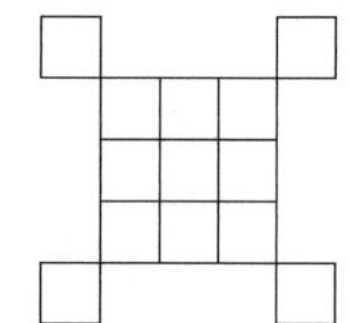

Linear/Non-linear

4 Consider the number of dots.

 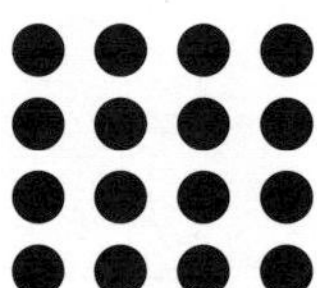 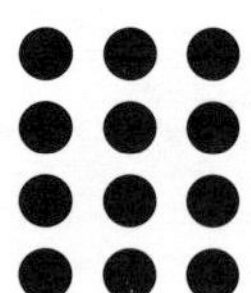 

Linear/Non-linear

5 100, 95, 90, 85, 80, …

Linear/Non-linear

6 1, 3, 6, 10, 15, …

Linear/Non-linear

ISBN: 9780170451505

Revision 1

1 Draw the next two shapes of this pattern.

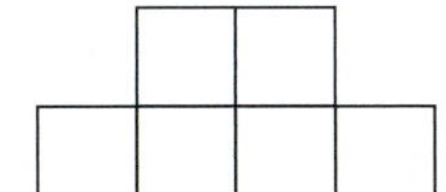

2 Complete the table for this pattern.

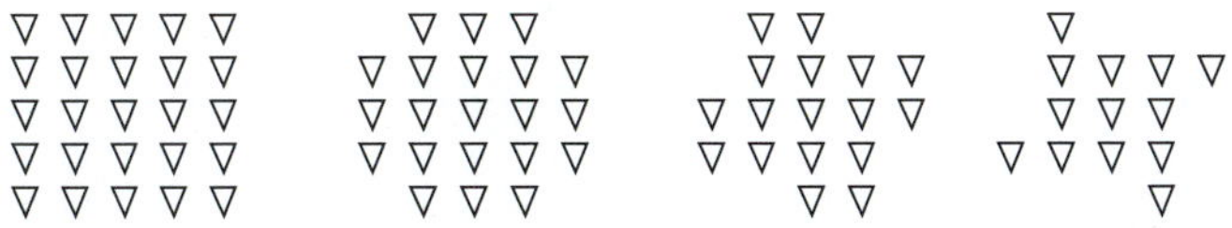

Shape number	Number of triangles
1	25
2	
3	
4	
5	
6	

3 **a** Complete the table.

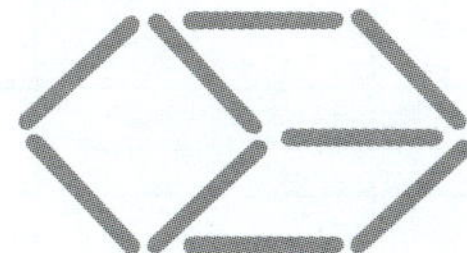

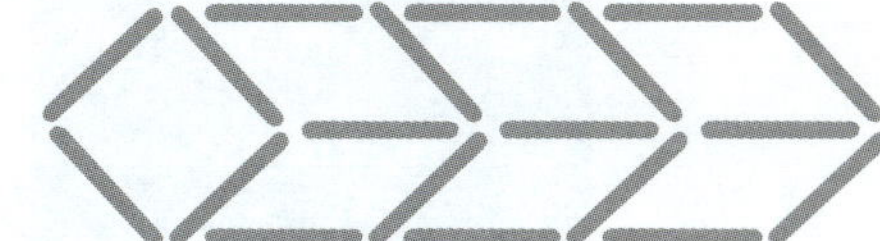

Term number (n)	0	1	2	3	4	5	6
Number of popsicle sticks (P)		9					

b Find the rule.

Number of popsicle sticks = ________ x term number + ________

P = ____________________

c How many popsicle sticks would be needed for the 50th shape?

P = ____________________

= ____________________

ISBN: 9780170451505

4 Complete the table and find the rule.

Term number (n)	0	1	2	3	4	5
Value of term (T)		3	5	7		

Rule:

$T =$ ______ $n +$ ______

5 Find the missing picture/values for these sequences.

a 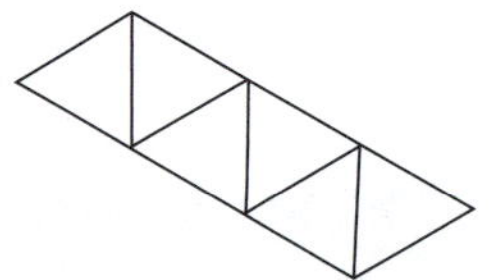 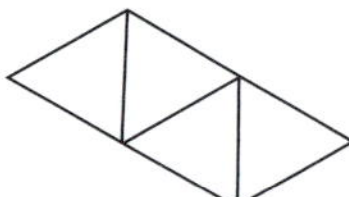 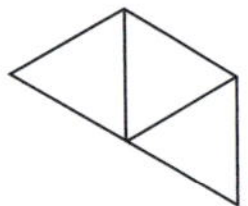

b ______, 7, ______, 13, ______, 19, …

6 Find the rules for these sequences.

a 7, 12, 17, 22, 27, … Rule __________ The 15th term = __________

b 21, 18, 15, 12, 9, … Rule __________ The 22nd term = __________

7 Use the rules to calculate the first five terms of these sequences.

a $V = 3n - 4$

Term number (n)	1	2	3	4	5
Calculations					
Value of term (V)					

Sequence: ____________________

b $y = -2x + 5$

Term number (x)	Calculations	Value of term (y)
1		
2		
3		
4		
5		

Sequence: ____________________

 ISBN: 9780170451505

8 Complete the table, plot the points, and then draw the line that represents the relationship $y = 2x + 1$.

x	Calculation	y	Coordinates
0			
1			
2			
3			
4			
5			
6			

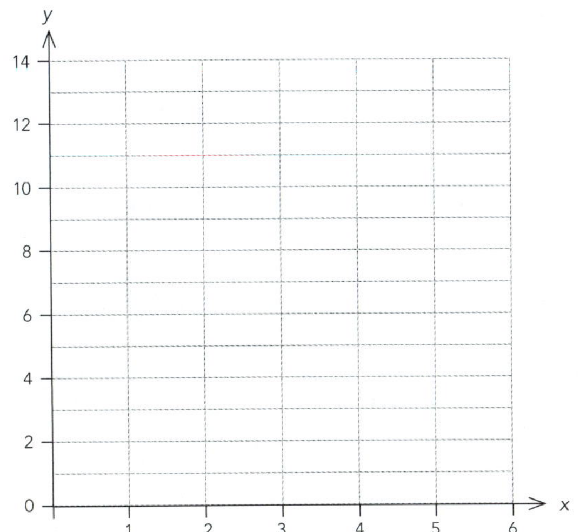

9 Ari has a 25-metre-long roll of wire netting. He is using it to make little fences to put around each of his trees to protect them from being eaten by hares. Each tree requires 2 m of netting.

a Complete the table to show how much netting he has left after protecting t trees.

Trees (t)	Remaining netting (m) (n)
1	
2	
3	
4	
5	

b Use the table to help you draw the points on the graph that shows how much netting he has left after he has protected t trees.

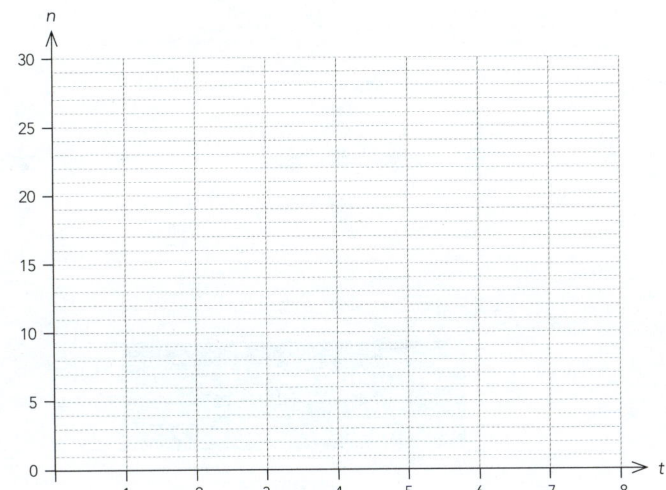

c Write an equation that relates how much netting he has left to the number of trees he has protected.

$n =$ ______________________

d Use the equation to work out much netting he will have left after he has protected 11 trees.

____

____

ISBN: 9780170451505

Revision 2

1 Draw the next two shapes of this pattern.

2 Complete the table for this pattern.

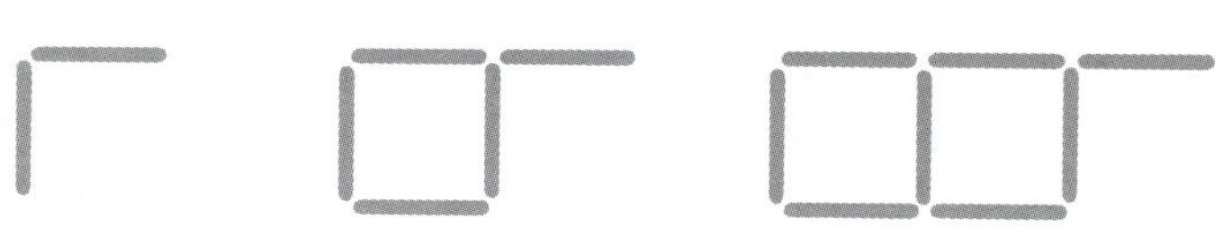

Shape number	Number of popsicle sticks
1	2
2	
3	
4	
5	
6	

3 **a** Complete the table.

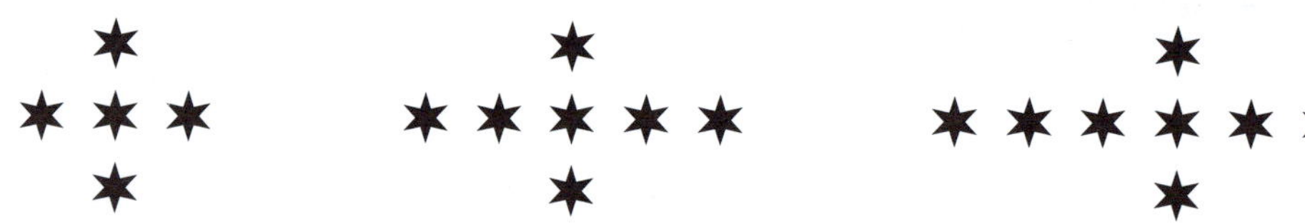

Term number (n)	0	1	2	3	4	5	6
Number of stars (S)		5					

b Find the rule.

Number of stars = ________ x term number + ________

S = ____________________

c How many stars would be needed for the 50th shape?

S = ____________________

= ____________________

 ISBN: 9780170451505

4 Complete the table and find the rule.

Term number (n)	0	1	2	3	4	5
Value of term (T)		5	8	11		

Rule:

$T =$ ______ $n +$ ______

5 Find the missing picture/values for these sequences.

a   

b ______, ______, 13, ______, 23, ______, 33, …

6 Find the rules for these sequences.

a 6, 9, 12, 15, 18, … Rule ______ The 21st term = ______

b 19, 17, 15, 13, 11, … Rule ______ The 15th term = ______

7 Use the rules to calculate the first five terms of these sequences.

a $V = 4n - 1$

Term number (n)	1	2	3	4	5
Calculations					
Value of term (V)					

Sequence: ______

b $y = -3x + 6$

Term number (x)	Calculations	Value of term (y)
1		
2		
3		
4		
5		

Sequence: ______

ISBN: 9780170451505

8 Complete the table, plot the points, and then draw the line that represents the relationship $y = -x + 10$.

x	Calculation	y	Coordinates
0			
1			
2			
3			
4			
5			
6			

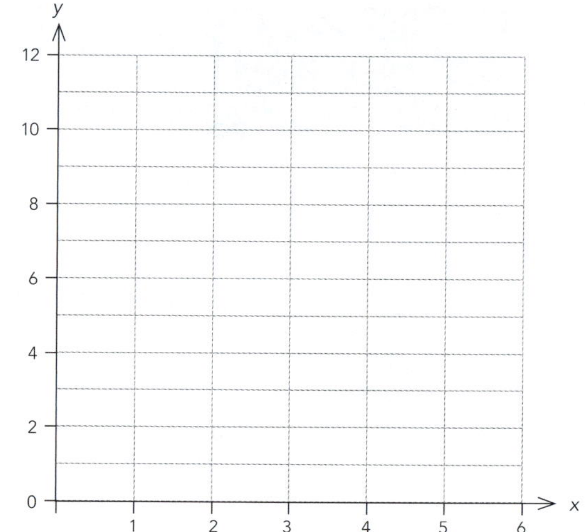

9 Tobias keeps chickens, and sells cartons of eggs for $7.50. However, he has to pay for the feed for the chickens, so his profit on each carton of eggs is $6. He has $18 in his piggy bank. He adds the profit from each carton to his piggy bank.

a Complete the table to show how much he has in his piggy bank (p) after selling c cartons of eggs.

Cartons (c)	Piggy bank ($) ($p$)
1	
2	
3	
4	
5	

b Use the table to help you draw the points on the graph that show how much he has in his piggy bank after he has sold c cartons of eggs.

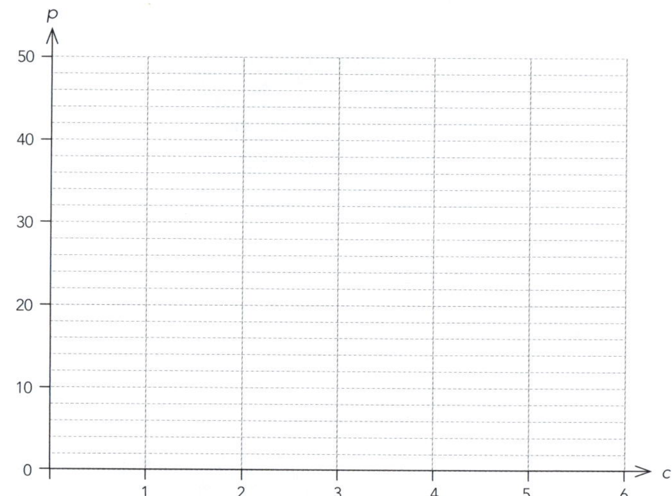

c Write an equation that relates the amount he has in his piggy bank to the number of cartons of eggs he has sold.

$p =$ ______________________

d Use the equation to work out much he will have in his piggy bank after he has sold 8 cartons of eggs.

____

____

 ISBN: 9780170451505

Answers

Substitution (p. 5)

Formula	$n = 3$	$n = 7$	$n = 10$
$A = n + 3$	$A = 6$	$A = 10$	$A = 13$
$A = 5n$	$A = 15$	$A = 35$	$A = 50$
$A = 6n - 4$	$A = 14$	$A = 38$	$A = 56$
$A = -2n + 1$	$A = -5$	$A = -13$	$A = -19$
$A = -3n - 1$	$A = -10$	$A = -22$	$A = -31$

Patterns (pp. 6–42)

Continuing patterns (p. 7)

1

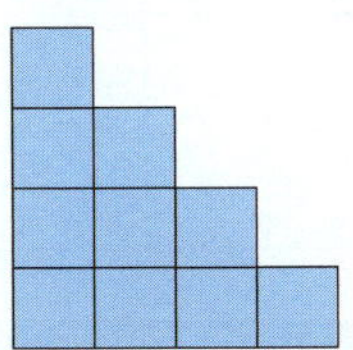

2

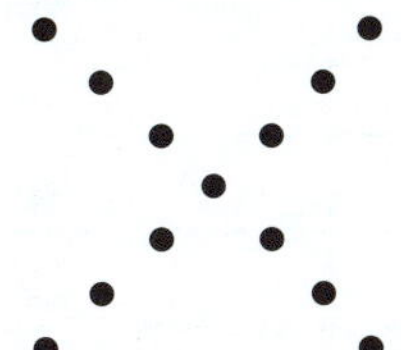

3

4

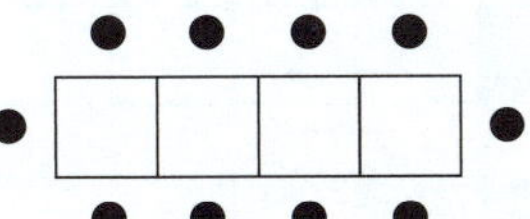

5

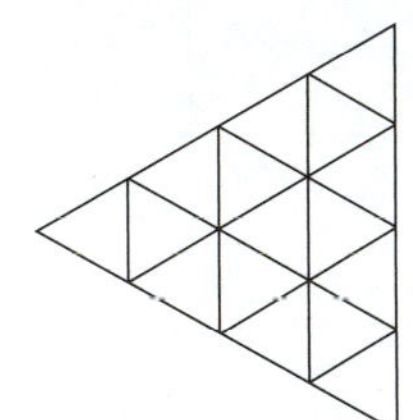

6

From pictures to numbers (p. 8)

1 2 4 6 8
2 10 7 4
3 1 3 5 7
4 5 9 13
5 22 15 8
6 1 3 5 7

Describing patterns (p. 9)

1 I started with **3 popsicle sticks** and then I added/~~subtracted~~ 2 each time.

2 I started with **1 dot** and then I added/~~subtracted~~ **4** each time.

3 I started with **16 stars** and then I ~~added~~/subtracted **2** each time.

4 I started with **1 square and then I added 4 each time**.

5 I started with **10 crosses and then I subtracted 2 each time**.

Challenge 1 (p.10)

1 The black blocks should not be there.

 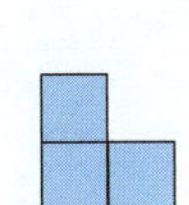 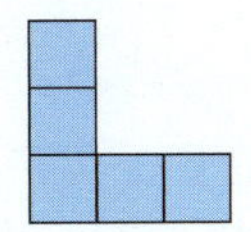 

2 The last stick (in blue) should not be there.

3 There should be three more dots (in blue) on the last shape.

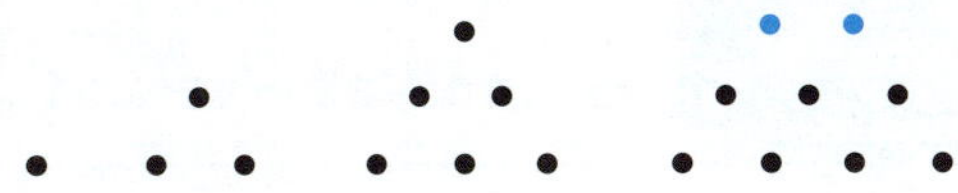

4 There is an extra stick (in blue) in the second shape.

5 1, 3, 5, **8**, 9, 11
The blue number should be 7.
6 72, 69, 66, 63, **61**, 57
The blue number should be 60.
7 **–2**, 1, 5, 9, 13, 17
The blue number should be –3.
8 10, 3, –4, **–10**, –18, –25
The blue number should be –11.

Patterns into tables (pp. 11–12)

1

Shape number	Number of crosses
1	6
2	8
3	10
4	12
5	14
6	16

+ 2 + 2 + 2 + 2 + 2

2

Shape number	1	2	3	4	5	6
Number of popsicle sticks	26	21	16	11	6	1

– 5 – 5 – 5

3

Shape number	Number of dots
1	4
2	7
3	10
4	13
5	16
6	19

+ 3

4

Shape number	1	2	3	4	5	6
Number of moons	15	14	13	12	11	10

– 1

Finding term 'zero' (pp. 13–14)

1

Term number	Number of squares
0	5
1	6
2	7
3	8
4	9
5	10

2

Term number	0	1	2	3	4	5
Number of crosses	12	10	8	6	4	2

3

Term number	0	1	2	3	4	5	6
Number of popsicle sticks	2	6	10	14	18	22	26

4

Term number	Number of popsicle sticks
0	46
1	38
2	30
3	22
4	14
5	6

5

Term number	0	1	2	3	4	5	6
Number of popsicle sticks	31	26	21	16	11	6	1

6

Term number	0	1	2	3	4	5	6
Number of popsicle sticks	20	28	36	44	52	60	68

ISBN: 9780170451505

Finding the rule from shapes and a table (pp. 15–26)

Increasing patterns

1

Term number (*n*)	Number of popsicle sticks (*P*)
0	1
1	3
2	5
3	7
4	9
5	11
6	13

In words: I started with 3 popsicle sticks and then I added/~~subtracted~~ 2 each time.
Find the mathematical rule:
Number of popsicle sticks
= 2 x term number + 1
Tidy it up: $P = 2n + 1$
Word rule: The number of popsicle sticks is calculated by multiplying the term number by 2 and then adding/~~subtracting~~ 1.
Use the rule:

$P = 2 \times 40 + 1$
$= 81$

2

Term number (*n*)	0	1	2	3	4	5	6
Number of popsicle sticks (*P*)	2	5	8	11	14	17	20

In words:
I started with 5 popsicle sticks and then I added/~~subtracted~~ 3 each time.
Find the mathematical rule:
Number of popsicle sticks
= 3 x term number + 2
Tidy it up: $P = 3n + 2$
Word rule: The number of popsicle sticks is calculated by multiplying the term number by 3 and then adding/~~subtracting~~ 2.
Use the rule:

$P = 3 \times 50 + 2$
$= 152$

3

Term number (*n*)	Number of dots (*D*)
0	–3
1	1
2	5
3	9
4	13
5	17
6	21

In words: I started with 1 dot and then I added/~~subtracted~~ 4 each time.
Find the mathematical rule:
Number of dots = 4 x term number – 3
Tidy it up: $D = 4n - 3$
Word rule: The number of dots is calculated by multiplying the term number by 4 and then ~~adding~~/subtracting 3.
Use the rule:

$D = 4 \times 40 - 3$
$= 157$

4

Term number (*n*)	0	1	2	3	4	5	6
Number of popsicle sticks (*P*)	0	6	12	18	24	30	36

In words: I started with 6 popsicle sticks and then I added/~~subtracted~~ 6 each time.
Find the mathematical rule:
Number of popsicle sticks = 6 x term number + 0
Tidy it up: $P = 6n + 0$ or $P = 6n$
Word rule: The number of popsicle sticks is calculated by multiplying the term number by 6 and then adding/~~subtracting~~ 0.
Use the rule:

$P = 6 \times 50$
$= 300$

Decreasing patterns

1

Term number (*n*)	Number of squares (*S*)
0	9
1	8
2	7
3	6
4	5
5	4

ISBN: 9780170451505

In words: I started with 8 squares and then I ~~added~~/subtracted 1 each time.
Find the mathematical rule:
Number of squares = –1 x term number + 9
Tidy it up: $S = -n + 9$
Word rule: The number of squares is calculated by multiplying the term number by –1 and then adding/~~subtracting~~ 9.
Use the rule:

$S = -6 + 9$
$= 3$

2

Term number (n)	0	1	2	3	4	5
Number of popsicle sticks (P)	22	19	16	13	10	7

In words: I started with 19 popsicle sticks and then I ~~added~~/subtracted 3 each time.
Find the mathematical rule:
Number of popsicle sticks
= –3 x term number + 22
Tidy it up: $P = -3n + 22$
Word rule: The number of popsicle sticks is calculated by multiplying the term number by –3 and then adding/~~subtracting~~ 22.
Use the rule:

$P = -3 \times 6 + 22$
$= 4$

3

Term number (n)	Number of snowflakes (S)
0	23
1	20
2	17
3	14
4	11
5	8

In words: I started with 20 snowflakes and then I ~~added~~/subtracted 3 each time.
Find the mathematical rule:
Number of snowflakes = –3 x term number + 23
Tidy it up: $S = -3n + 23$
Word rule: The number of snowflakes is calculated by multiplying the term number by –3 and then adding/~~subtracting~~ 23.
Use the rule:

$S = -3 \times 7 + 23$
$= 2$

4

Term number (n)	0	1	2	3	4	5
Number of popsicle sticks (P)	66	60	54	48	42	36

In words: I started with 60 popsicle sticks and then I ~~added~~/subtracted 6 each time.
Find the mathematical rule:
Number of popsicle sticks
= –6 x term number + 66
Tidy it up: $P = -6n + 66$
Word rule: The number of popsicle sticks is calculated by multiplying the term number by –6 and then adding/~~subtracting~~ 66.
Use the rule:

$P = -6 \times 10 + 66$
$= 6$

Challenge 2 (p. 27)

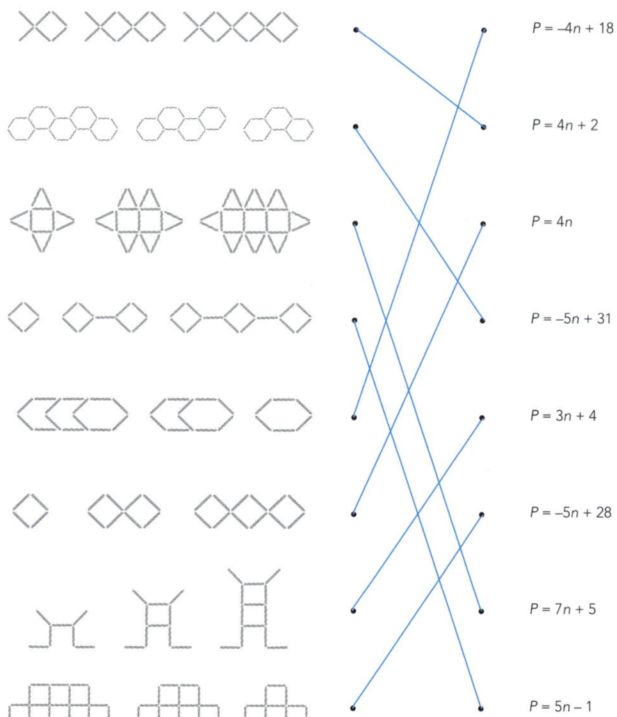

ISBN: 9780170451505

Finding the rule from a table (pp. 28–30)

1

Term number (n)	Value of term (T)
0	4
1	6
2	8
3	10
4	12

+ 2

Rule: $T = 2n + 4$

2

Term number (n)	0	1	2	3	4	5
Value of term (T)	3	6	9	12	15	18

+ 3

Rule: $T = 3n + 3$

3

Term number (n)	0	1	2	3	4	5
Value of term (T)	0	4	8	12	16	20

Rule: $T = 4n + 0$ or $T = 4n$

4

Term number (n)	Value of term (T)
0	0
1	10
2	20
3	30
4	40

Rule: $T = 10n + 0$ or $T = 10n$

5

Term number (n)	0	1	2	3	4	5
Value of term (T)	21	19	17	15	13	11

Rule: $T = -2n + 21$

6

Term number (n)	Value of term (T)
0	55
1	50
2	45
3	40
4	35

Rule: $T = -5n + 55$

7

Term number (n)	1	2	3	4	5
Value of term (T)	4	10	16	22	28

Rule: $T = 6n - 2$

8

Term number (n)	1	2	3	4	5
Value of term (T)	2	–1	–4	–7	–10

Rule: $T = -3n + 5$

9

Term number (n)	Value of term (T)
1	100
2	96
3	92
4	88

Rule: $T = -4n + 104$

10

Term number (n)	1	2	3	4	5
Value of term (T)	–4	–6	–8	–10	–12

Rule: $T = -2n - 2$

Challenge 3 (p. 31)

1

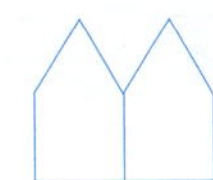
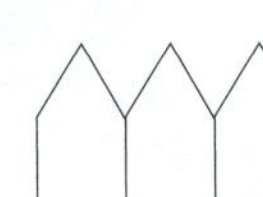
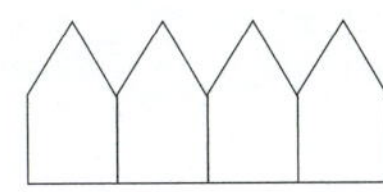

2

3

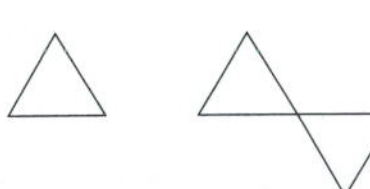
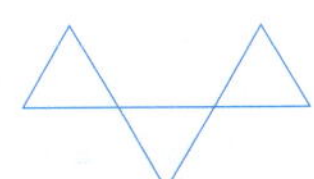
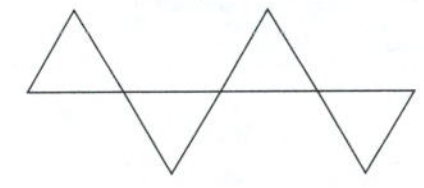

ISBN: 9780170451505

4 6, 12, 18, **24**, 30, 36
5 7, 9, 11, 13, **15**, 17, **19**
6 12, **24**, 36, **48**, 60, 72
7 2, 5, **8**, **11**, 14, **17**
8 **28**, 34, 40, **46**, 52, **58**
9 **84**, 75, **66**, 57, **48**, 39, **30**
10 81, **75**, **69**, **63**, **57**, 51

Finding the rule from a list (pp. 32–33)

1	$T = 6n + 2$	2	$T = 3n + 8$
3	$T = 4n - 1$	4	$T = -2n + 18$
5	$T = -5n + 35$	6	$T = 10n - 10$
7	$T = -3n + 9$	8	$T = 12n - 24$

9 4, 7, 10, 13, …
−3 +3 +3 +3
I started with **4** and then I added/~~subtracted~~ **3** each time.
Term zero will be **1**.
Rule: $T =$ **$3n + 1$**

10 I started with **1** and then I added/~~subtracted~~ **2** each time.
Term zero will be **−1**.
Rule: $T =$ **$2n - 1$**

11 I started with **24** and then I ~~added~~/subtracted **4** each time.
Term zero will be **28**.
Rule: $T =$ **$-4n + 28$**

12 I started with **−5** and then I added/~~subtracted~~ **5** each time.
Term zero will be **−10**.
Rule: $T =$ **$5n - 10$**

13 I started with **20** and then I ~~added~~/subtracted **6** each time.
Term zero will be **26**.
Rule: $T =$ **$-6n + 26$**

14 Rule: $T =$ **$10n + 2$**
15 Rule: $T =$ **$-2n + 8$**
16 Rule: $T =$ **$3n + 10$**
17 Rule: $T =$ **$9n + 0$** or $T =$ **$9n$**

Finding a term from a rule (p. 34)

1	101	2	115
3	1100	4	−94
5	38	6	−10
7	10	8	−20
9	−13	10	−120

Finding the sequence from the rule (pp. 35–36)

1 $T = 5n + 2$

Term number (n)	1	2	3	4	5
Calculations	5 x **1** + 2				**5 x 5 + 2**
Value of term (T)	**7**	**12**	**17**	**22**	**27**

Sequence: **7, 12, 17, 22, 27**

2 $T = -3n + 10$

Term number (n)	1	2	3	4	5
Calculations	**−3 x 1 + 10**				**−3 x 5 + 10**
Value of term (T)	**7**	**4**	**1**	**−2**	**−5**

Sequence: **7, 4, 1, −2, −5**

3 $T = 6n - 1$

Term number (n)	Calculations	Value of term (T)
1	6 x **1** − 1	**5**
2		**11**
3		**17**
4		**23**
5	**6 x 5 − 1**	**29**

Sequence: **5, 11, 17, 23, 29**

4 $T = 4n - 11$

Term number (n)	1	2	3	4	5
Calculations	**4 x 1 − 11**				**4 x 5 − 11**
Value of term (T)	**−7**	**−3**	**1**	**5**	**9**

Sequence: **−7, −3, 1, 5, 9**

5 $T = 10n - 100$

Term number (n)	1	2	3	4	5
Calculations	**10 x 1 − 100**				**10 x 5 − 100**
Value of term (T)	**−90**	**−80**	**−70**	**−60**	**−50**

Sequence: **−90, −80, −70, −60, −50**

6 $T = -5n - 2$

Term number (n)	1	2	3	4	5
Calculations	**−5 x 1 − 2**				**−5 x 5 − 2**
Value of term (T)	**−7**	**−12**	**−17**	**−22**	**−27**

Sequence: **−7, −12, −17, −22, −27**

Applications (p. 37)

1 a He started with \$75 − \$15 = \$60
b Term zero
c \$75, \$90, \$105, …
d $T = 15n + 60$
e $T = 15 \times 11 + 60$
= \$225

ISBN: 9780170451505

2 **a** \$21, \$32, \$43, …
b Term zero = \$10 **c** $T = 11n + 10$
d $T = 11 \times 8 + 10$
$= \$98$

3 **a** \$24, \$32, \$40, …
b Term zero = \$16 **c** $T = 8n + 16$
d $T = 8 \times 9 + 16$
$= \$88$

4 **a** 35, 32, 29, …
b Term zero = 38 **c** $T = -3n + 38$
d $T = -3 \times 11 + 38$
$= 5$

5 **a** \$334, \$369, \$404, …
b Term zero = \$299 **c** $T = 35n + 299$
d $T = 35 \times 12 + 299$
$= \$719$

6 **a** 92, 84, 76, …
b Term zero = 100 **c** $T = -8n + 100$
d $T = -8 \times 9 + 100$
$= 18$

7 **a** –5°, –7°, –9°, …
b Term zero = –3° **c** $T = -2n - 3$
d $T = -2 \times 8 - 3$
$= -19°$

Challenge 4 (p. 40)

1

2	**4**	**6**	8
5	**10**	15	**20**
8	16	**24**	**32**
11	**22**	33	**44**

2

16	**17**	18	**19**
25	35	**45**	**55**
34	**53**	72	**91**
43	**71**	**99**	**127**

3

11	9	**7**	5
19	**18**	**17**	**16**
27	**27**	27	**27**
35	**36**	**37**	**38**

4

–30	**0**	30	**60**
–11	7	**25**	**43**
8	**14**	**20**	26
27	**21**	15	**9**

5

–1	5	**11**	**17**
7	**9**	11	**13**
15	**13**	**11**	9
23	17	**11**	**5**

6

–12	**–8**	**–4**	0
–15	**1**	17	**33**
–18	**10**	**38**	**66**
–21	**19**	**59**	99

7

–1	**10**	21	**32**
–6	**3**	**12**	**21**
–11	**-4**	**3**	**10**
–16	**–11**	–6	–1

8

–11	**4**	**19**	**34**
8	0	**–8**	**–16**
27	**–4**	**–35**	**–66**
46	**–8**	**–62**	–116

Cross-number (p. 41)

[1] 2	5	8	[2] 11		[3] 23	27	31	[4] 35
8			13		24			34
14		[5] 10	15	20	25	[6] 30		33
[7] 20	19	18	17		[8] 26	28	30	32
		26				26		
[9] 30	32	34	[10] 36		[11] 20	24	28	[12] 32
31		[13] 42	37	32	27	22		28
32			38		34			28
[14] 33	35	37	39		[15] 41	36	31	26

Challenge 5 (p. 42)

1

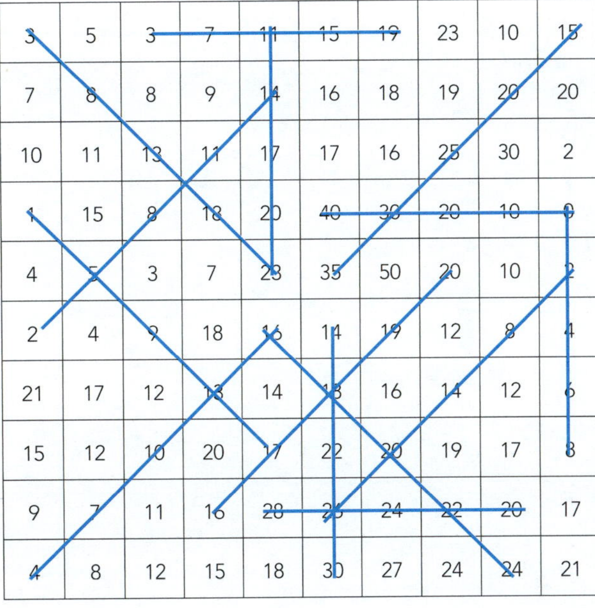

3	5	3	7	11	15	19	23	10	15
7	8	8	9	14	16	18	19	20	20
10	11	13	11	17	17	16	25	30	2
1	15	8	18	20	40	30	20	10	0
4	5	3	7	23	35	50	20	10	2
2	4	9	18	16	14	19	12	8	4
21	17	12	13	14	18	16	14	12	6
15	12	10	20	17	22	20	19	17	8
9	7	11	16	28	26	24	22	20	17
4	8	12	15	18	30	27	24	24	21

2 **a** 22, 24, 28, 34, 42, 52, **64, 78, 94**
+ 2 + 4 + 6 + 8 + 10 + 12 + 14 + 16
b 70, 68, 65, 61, 56, 50, **43**, **35**, **26**
c 3, 4, 8, 17, 33, 58, **94**, **143**, **207**

ISBN: 9780170451505

Graphs (pp. 43–78)

Plotting points (pp. 43–47)

Positive coordinates

1 **a** and **c**

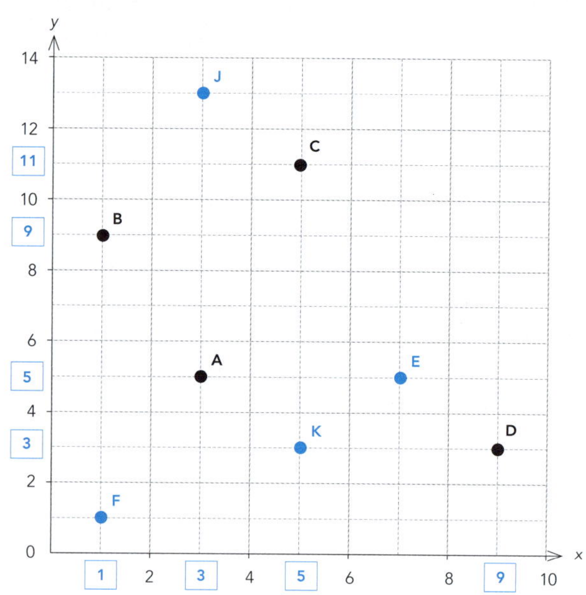

b A (3, 5) B (1, 9) C (5, 11) D (9, 3)

2 **a** and **c**

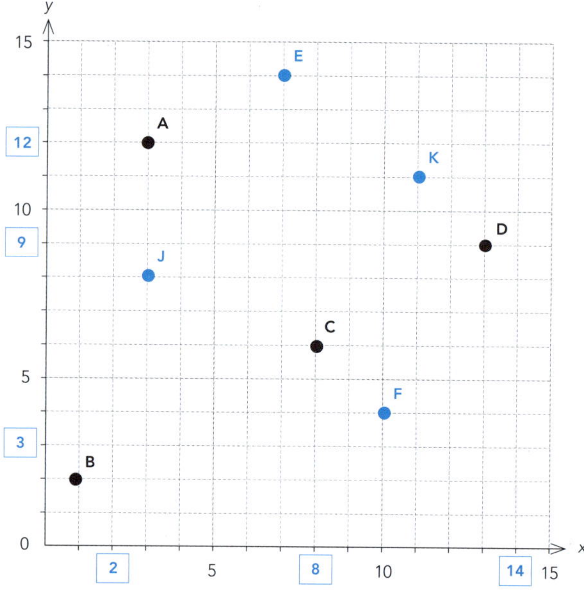

b A (3, 12) B (1, 2) C (8, 6) D (13, 9)

Negative coordinates

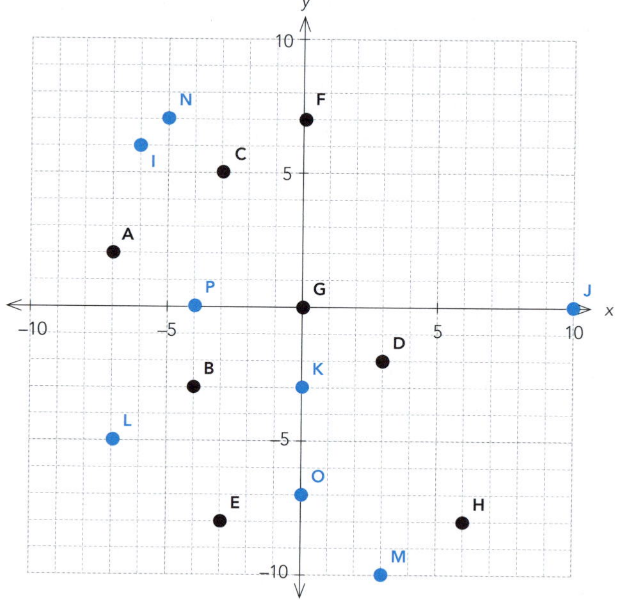

A (−6, 2) **B** (−4, −3)
C (−3, 5) **D** (3, −2)
E (−3, −8) **F** (0, 7)
G (0, 0) **H** (6, −8)

Given pattern — fill in the table, find the equation and plot the points (pp. 48–53)

1

Term number (x)	Value of the term (y)
1	5
2	9
3	13
4	17
5	21

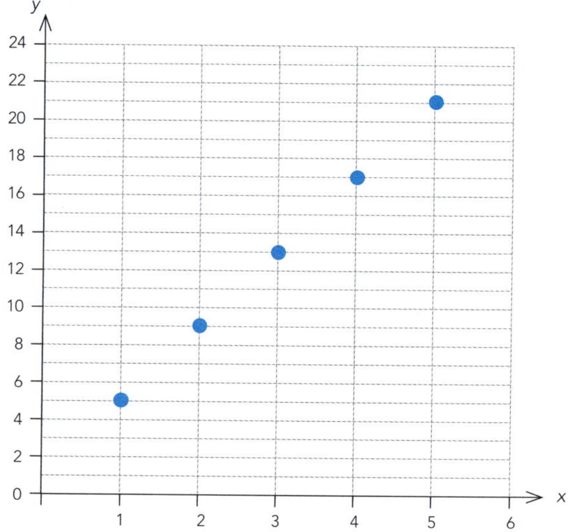

y = 4 x term number + 1
Tidy it up: $y = 4x + 1$

ISBN: 9780170451505

2

Term number (x)	Value of the term (y)
1	7
2	**10**
3	**13**
4	**16**
5	**19**

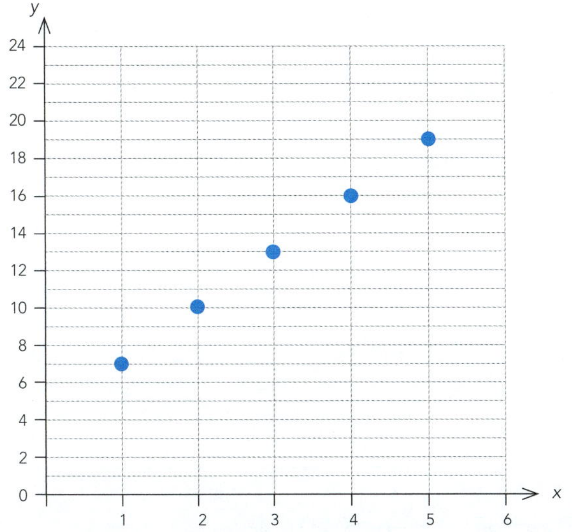

$y = 3$ x term number + 4

Tidy it up: $y = 3x + 4$

3

Term number (x)	Value of the term (y)
1	6
2	**11**
3	**16**
4	**21**
5	**26**

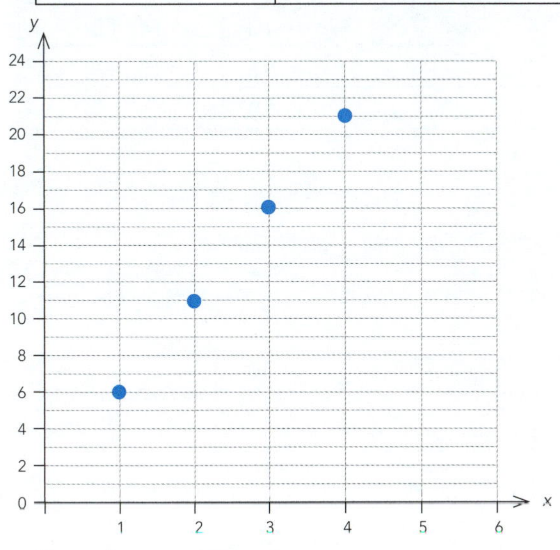

Equation: $y = 5x + 1$

4

Term number (x)	Value of the term (y)
1	10
2	**8**
3	**6**
4	**4**
5	**2**

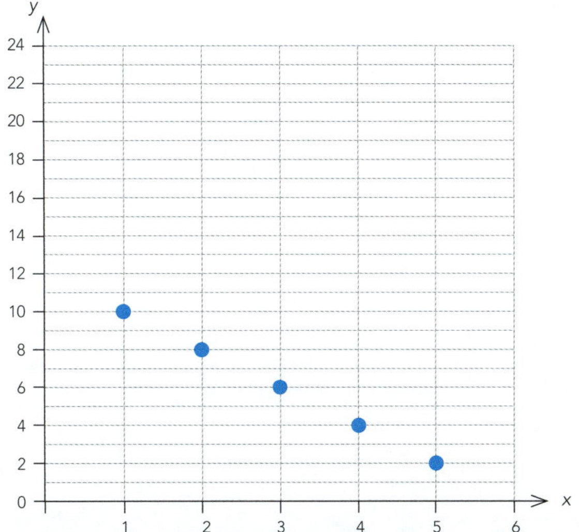

Equation: $y = -2x + 12$

5

Term number (x)	Value of the term (y)
1	21
2	**18**
3	**15**
4	**12**
5	**9**

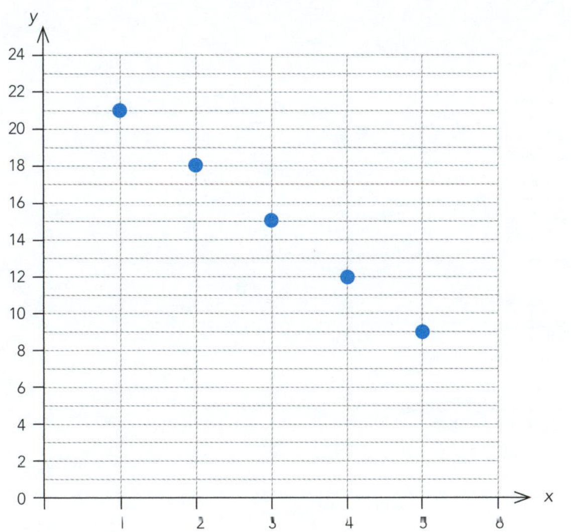

Equation: $y = -3x + 24$

ISBN: 9780170451505

6

Bags (x)	Money (y)
1	$8
2	$11
3	$14
4	$17
5	$20

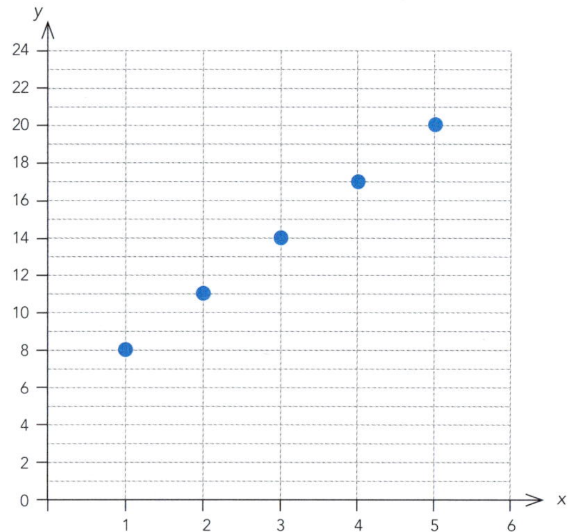

Equation: $y = 3x + 5$

$35

7

Ride (x)	Credit (y)
1	18
2	16
3	14
4	12
5	10

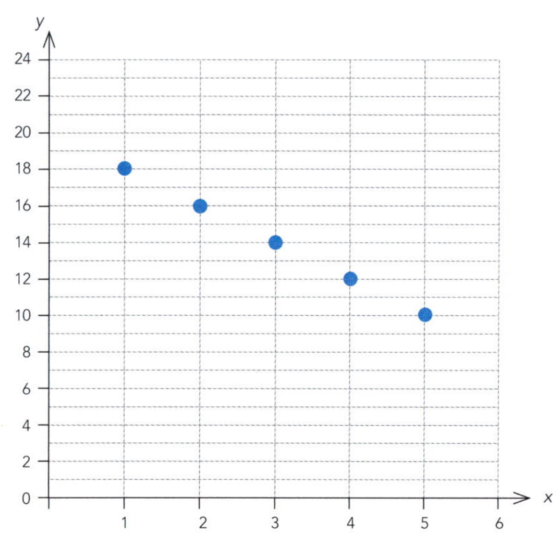

Equation: $y = -2x + 20$

$2

8

Steps (x)	Screws (y)
1	44
2	38
3	32
4	26
5	20

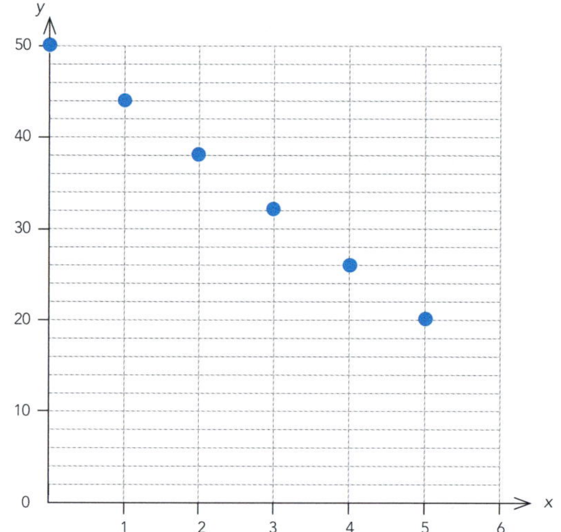

Equation: $y = -6x + 50$

2 screws left over

Given the equation — fill in the table and plot the points (pp. 54–57)

1

x	Calculation	y	Coordinates
0	3 x 0 + 1	1	(0, 1)
1		4	(1, 4)
2		7	(2, 7)
3		10	(3, 10)
4		13	(4, 13)
5		16	(5, 16)
6	3 x 6 + 1	19	(6, 19)

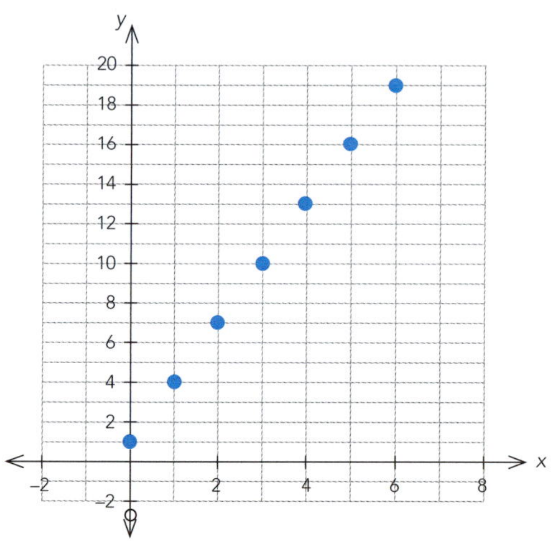

 ISBN: 9780170451505

2

x	Calculation	y	Coordinates
0	2 x 0 + 4	4	(0, 4)
1		6	(1, 6)
2		8	(2, 8)
3		10	(3, 10)
4		12	(4, 12)
5		14	(5, 14)
6	2 x 6 + 4	16	(6, 16)

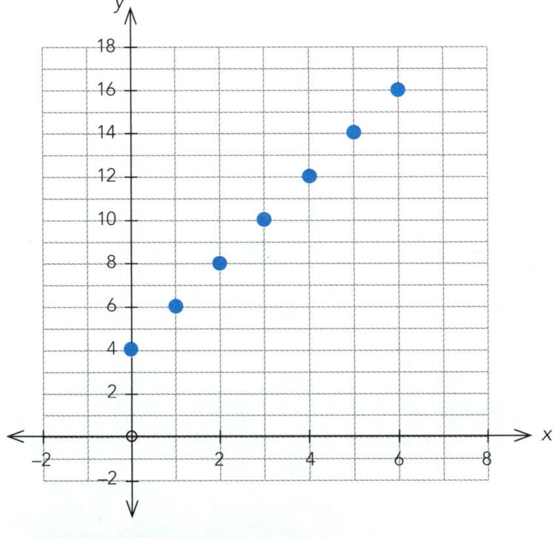

3

x	y
0	–2
1	2
2	6
3	10
4	14
5	18
6	22

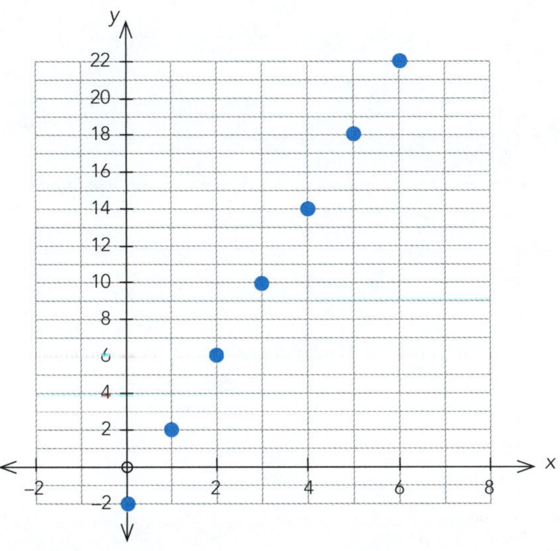

4

x	y
0	–1
1	4
2	9
3	14
4	19
5	24
6	29

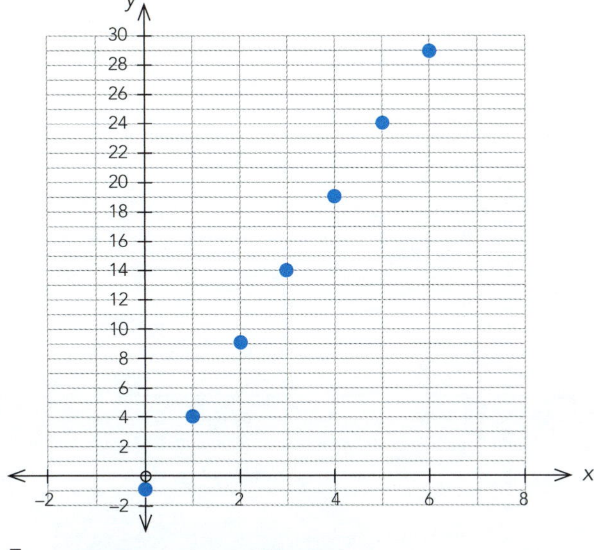

5

x	y
0	8
1	6
2	4
3	2
4	0
5	–2
6	–4

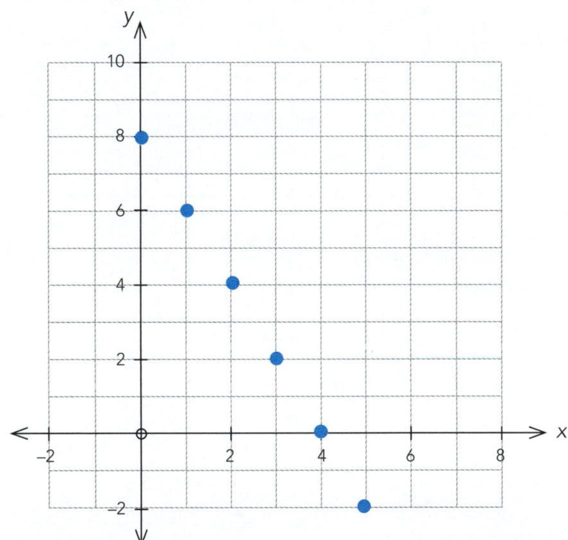

ISBN: 9780170451505

6

x	y
0	7
1	6
2	5
3	4
4	3
5	2
6	1

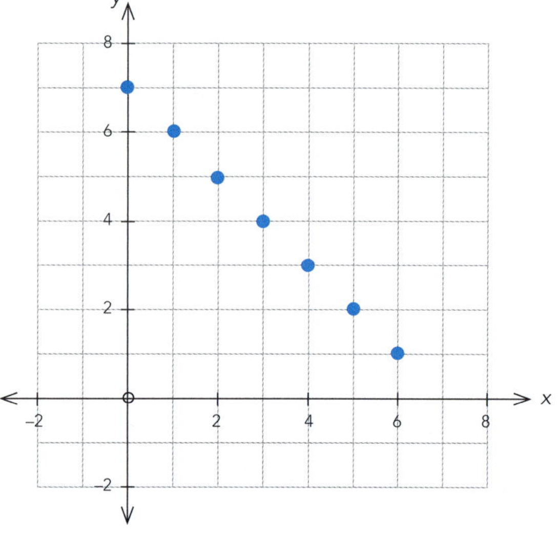

7

x	y
0	–3
1	–4
2	–5
3	–6
4	–7
5	–8
6	–9

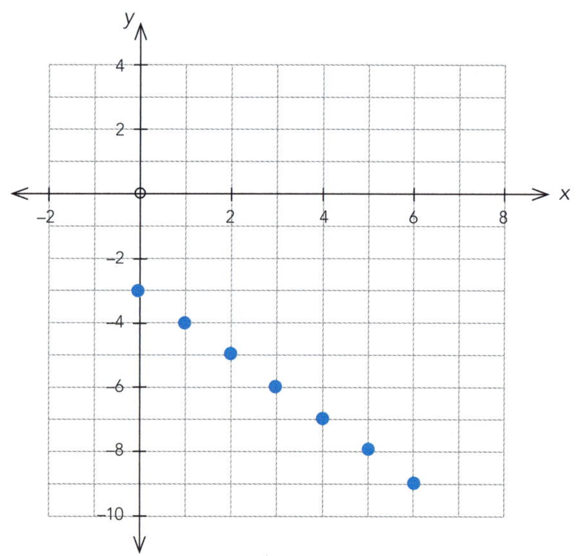

8

x	y
0	0
1	2
2	4
3	6
4	8
5	10
6	12

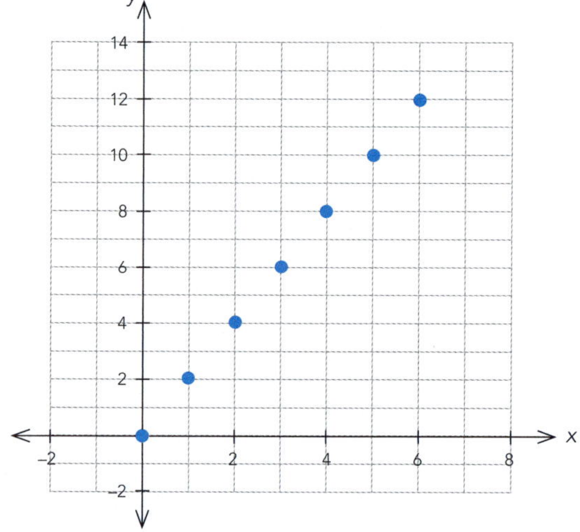

9

x	y
0	0
1	–1
2	–2
3	–3
4	–4
5	–5
6	–6

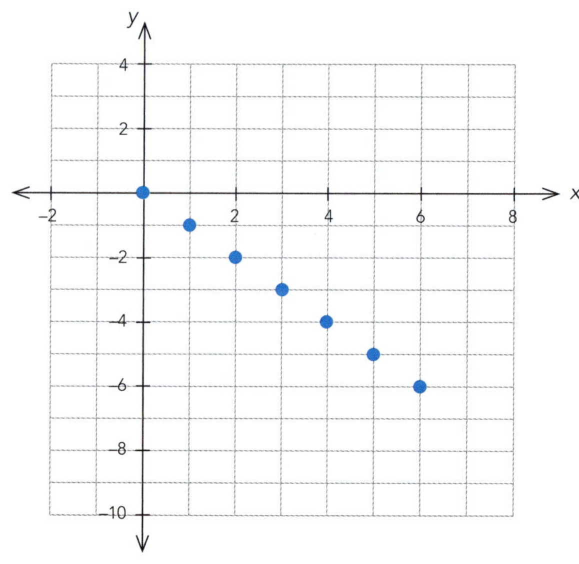

ISBN: 9780170451505

Given graph — fill in the table and write the equation (pp. 58–59)

1

x	y
0	4
1	5
2	6
3	7
4	8
5	9
6	10

Equation: $y = 1x + 4$ or $y = x + 4$

2

x	y
0	1
1	4
2	7
3	10
4	13
5	16
6	19

Equation: $y = 3x + 1$

3

x	y
0	0
1	5
2	10
3	15
4	20

Equation: $y = 5x + 0$ or $y = 5x$

4

x	y
0	9
1	7
2	5
3	3
4	1

Equation: $y = -2x + 9$

Applications (pp. 60–62)

1 a

Number of tickets (t)	Cost (c)
1	7
2	12
3	17
4	22
5	27

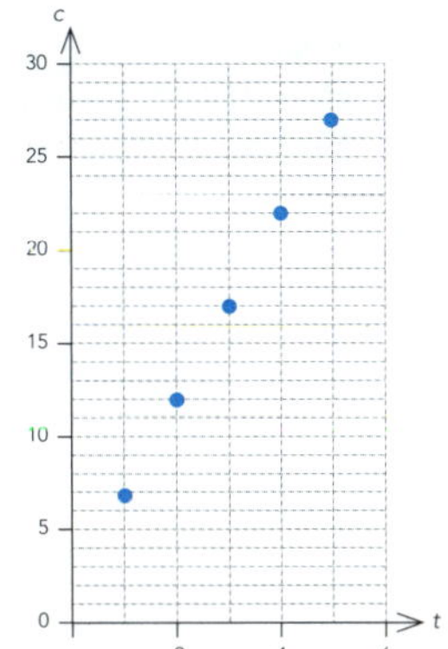

b $c = 5t + 2$

c $37

d 12 (with $3 left in her account)

2 a

Number of mangoes (m)	Cost (c)
1	4
2	7
3	10
4	13
5	16

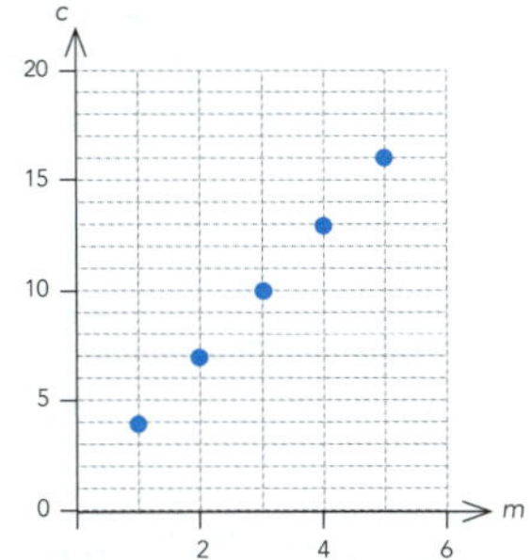

b $c = 3m + 1$

c $28

3 a

Number of jackets (j)	Number of buttons remaining (b)
1	26
2	22
3	18
4	14
5	10

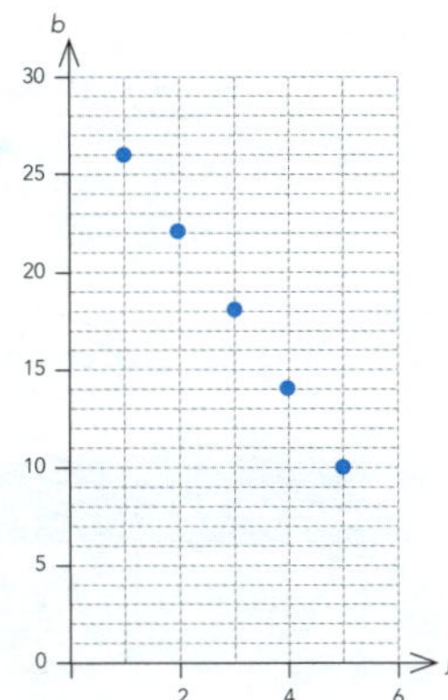

b $b = -4j + 30$

c $b = -4 \times 7 + 30 = 2$

Therefore she has 2 buttons remaining.

d 6 buttons

4 a

Time (hours) (h)	Cost ($) ($c$)
1	14
2	18
3	22
4	26
5	30

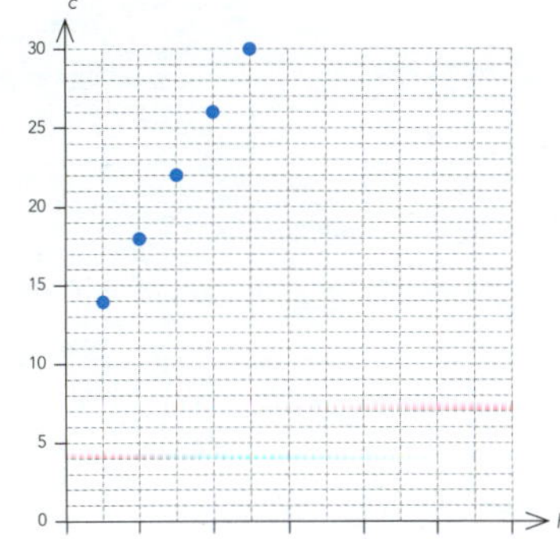

b $c = 4h + 10$

c $42

ISBN: 9780170451505

5 a

Distance (km) (d)	Cost (c)
2	10
4	16
6	22
8	28
10	34
12	40

b $c = 3d + 4$

c Chloe's Cabs charge a flat $4 fee plus $3 for each kilometre.

d

Distance (km) (d)	Cost (c)
2	14
4	18
6	22
8	26
10	30
12	34
14	38

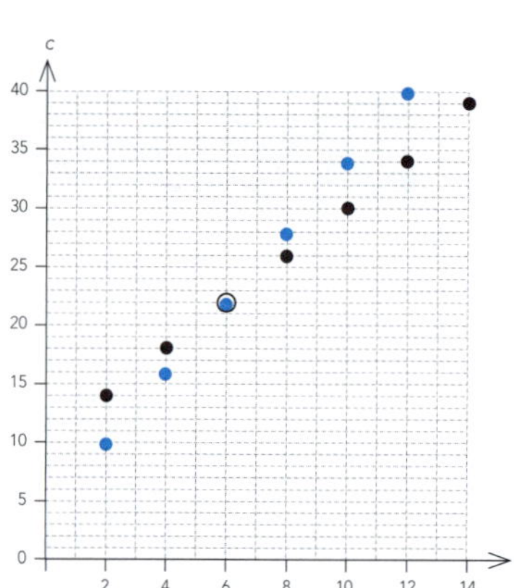

e $c = 2d + 10$

f $c = 2 \times 21 + 10$
$= \$52$

g A ride of 6 km costs $22 from both companies.

Drawing straight lines (pp. 63–66)

1

Term (x)	$3x + 2$	Answer (y)	Point
0	3 x 0 + 2	2	(0, 2)
1		5	(1, 5)
2		8	(2, 8)
3		11	(3, 11)
4	3 x 4 + 2	14	(4, 14)

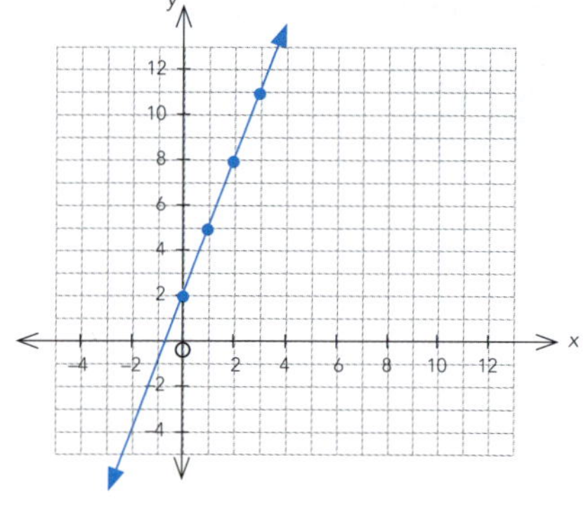

2

Term (x)	$2x + 1$	Answer (y)	Point
0	2 x 0 + 1	1	(0, 1)
1		3	(1, 3)
2		5	(2, 5)
3		7	(3, 7)
4	2 x 4 + 1	9	(4, 9)

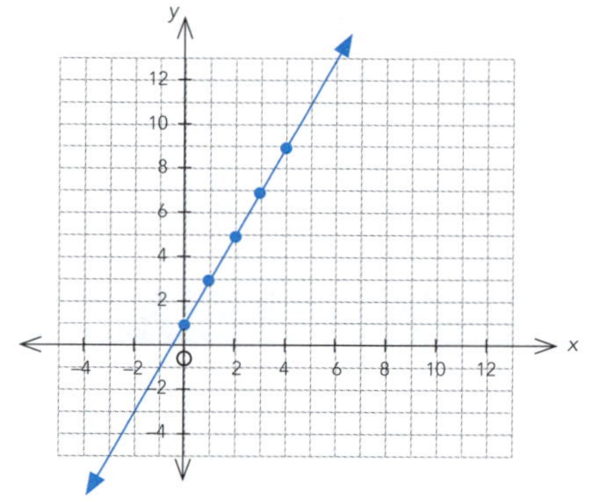

3

Term (x)	$x + 4$	Answer (y)	Point
0	0 + 4	4	(0, 4)
1		5	(1, 5)
2		6	(2, 6)
3		7	(3, 7)
4	4 + 4	8	(4, 8)

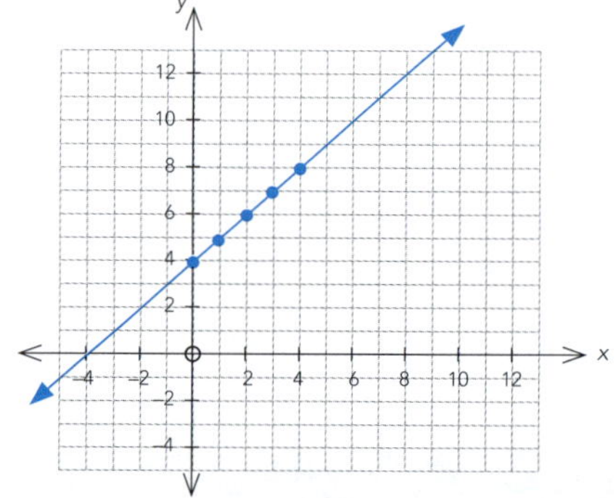

4

Term (x)	Answer (y)	Point
0	6	(0, 6)
1	4	(1, 4)
2	2	(2, 2)
3	0	(3, 0)
4	–2	(4, –2)

ISBN: 9780170451505

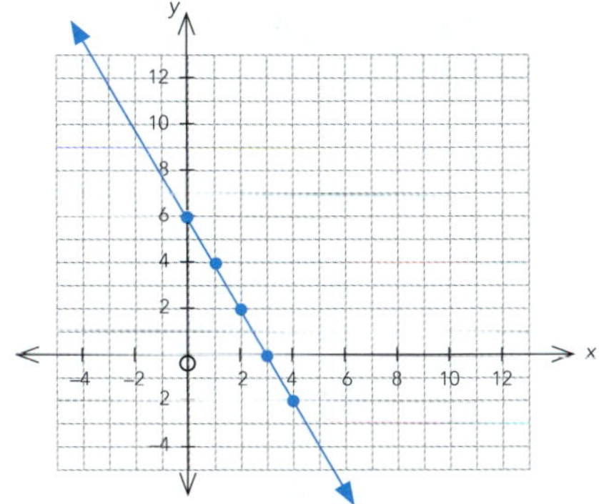

5

Term (x)	Answer (y)	Point
0	8	(0, 8)
1	7	(1, 7)
2	6	(2, 6)
3	5	(3, 5)
4	4	(4, 4)

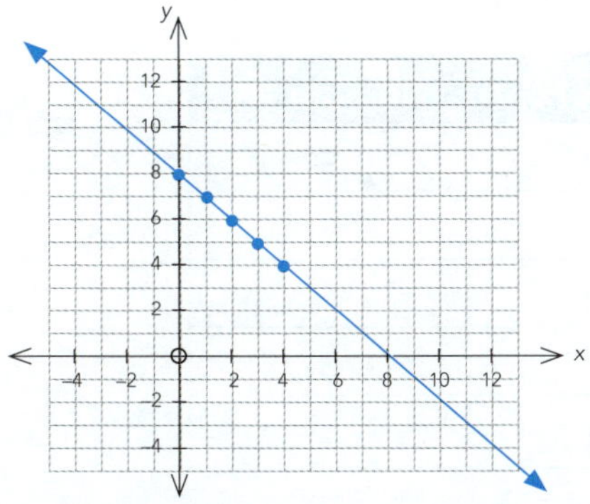

6

Term (x)	Answer (y)	Point
0	0	(0, 0)
1	4	(1, 4)
2	8	(2, 8)
3	12	(3, 12)
4	16	(4, 16)

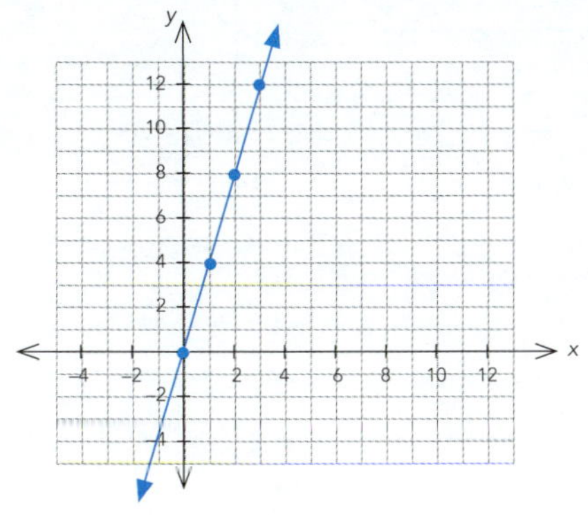

7

Term (x)	Answer (y)	Point
0	–3	(0, –3)
1	–1	(1, –1)
2	1	(2, 1)
3	3	(3, 3)
4	5	(4, 5)

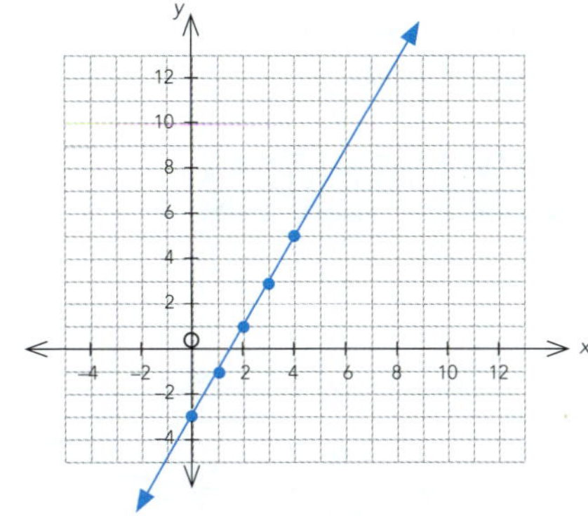

8

Term (x)	Answer (y)	Point
0	–4	(0, –4)
1	–3	(1, –3)
2	–2	(2, –2)
3	–1	(3, –1)
4	0	(4, 0)

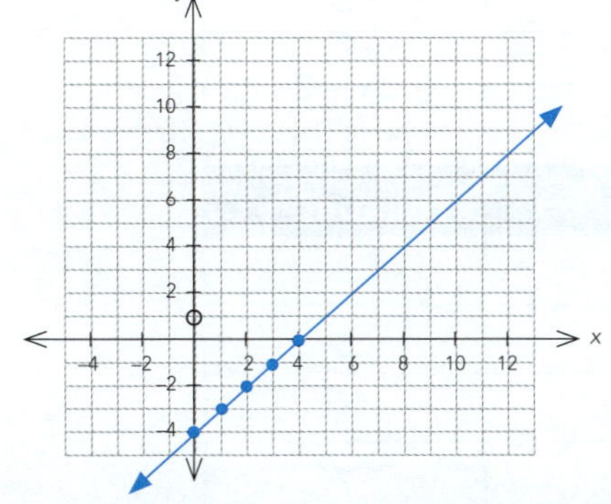

9

Term (x)	Answer (y)	Point
0	5	(0, 5)
1	4	(1, 4)
2	3	(2, 3)
3	2	(3, 2)
4	1	(4, 1)

ISBN: 9780170451505

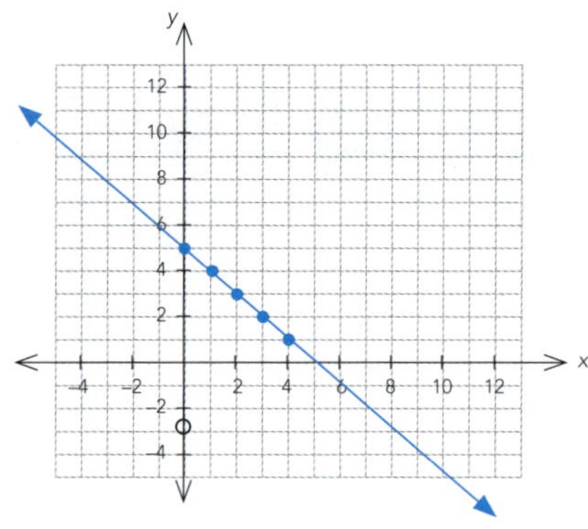

Given graph, write down the coordinates and find the equation (pp. 67–69)

1

Term (x)	Number (y)
0	5
1	6
2	**7**
3	**8**
4	**9**

The equation is y = **x + 5**

2

Term (x)	Number (y)
0	–3
1	**–1**
2	**1**
3	**3**
4	**5**

The equation is y = **2x – 3**

3

Term (x)	Number (y)
0	**4**
1	**5**
2	**6**
3	**7**
4	**8**

The equation is y = **x + 4**

4

Term (x)	Number (y)
0	**0**
1	**3**
2	**6**
3	**9**
4	**12**

The equation is y = **3x + 0** or y = **3x**

5

Term (x)	Number (y)
0	**–2**
1	**2**
2	**6**
3	**10**
4	**14**

The equation is y = **4x – 2**

6

Term (x)	Number (y)
0	**7**
1	**6**
2	**5**
3	**4**
4	**3**

The equation is y = **–x + 7**

7

Term (x)	Number (y)
0	**9**
1	**7**
2	**5**
3	**3**
4	**1**

The equation is y = **–2x + 9**

Horizontal and vertical lines (pp. 70–72)

1 $y = 3$ **2** $y = -1$

3 $y = 6$ **4** $x = 9$

5

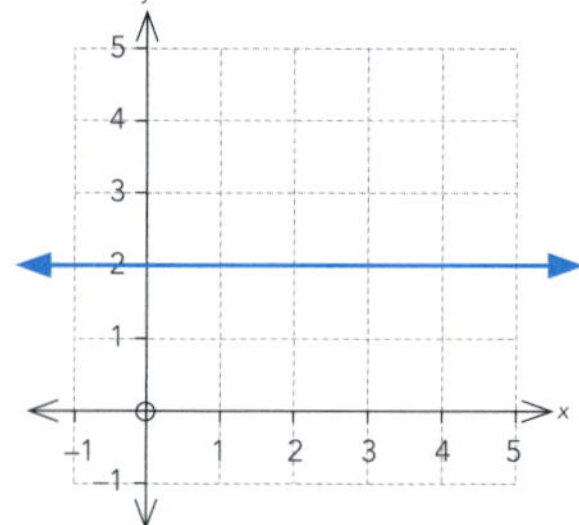

6

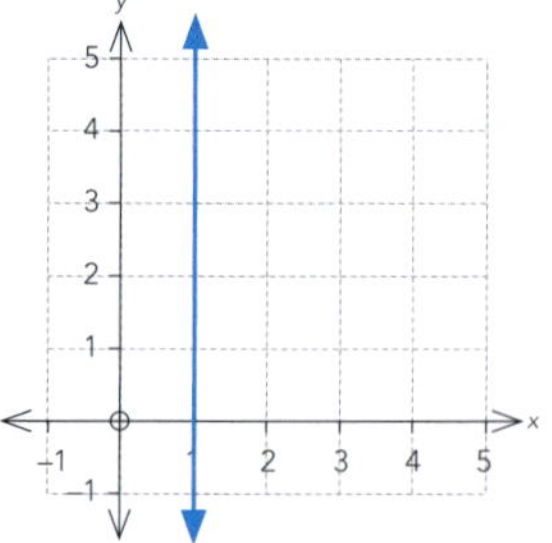

7

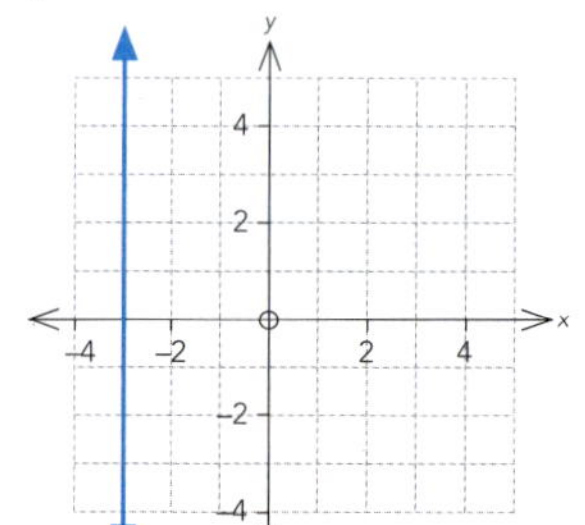

8

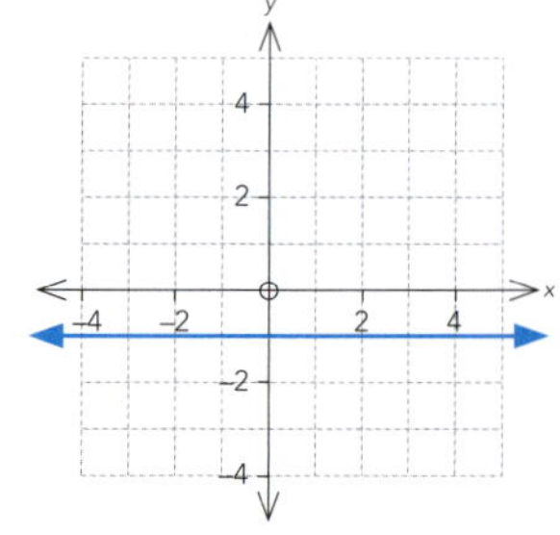

 ISBN: 9780170451505

9

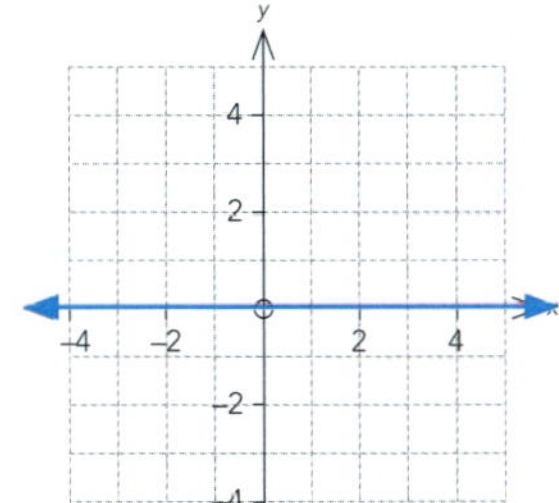

10

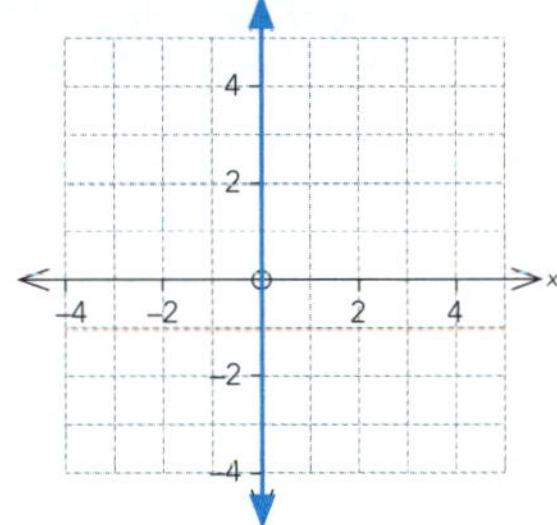

Applications (pp. 73–76)

1 **a**

Time (min) (t)	Depth (cm) (d)
0	14
1	22
2	30
3	38
4	46
5	54

b $d = 8t + 14$ **c** 14 cm

d 8 cm per minute

2 **a**

Week (w)	Saved ($) ($s$)
0	50
1	80
2	110
3	140
4	170
5	200

b $s = 30w + 50$ **c** $320

3 **a**

Week (w)	Owes ($) ($o$)
0	260
1	240
2	220
3	200
4	180
5	160

b $o = -20w + 260$ **c** $60

d $20

4 **a**

Time (min) (t)	Distance from school (d)
0	1800
1	1750
2	1700
3	1650
4	1600
5	1550

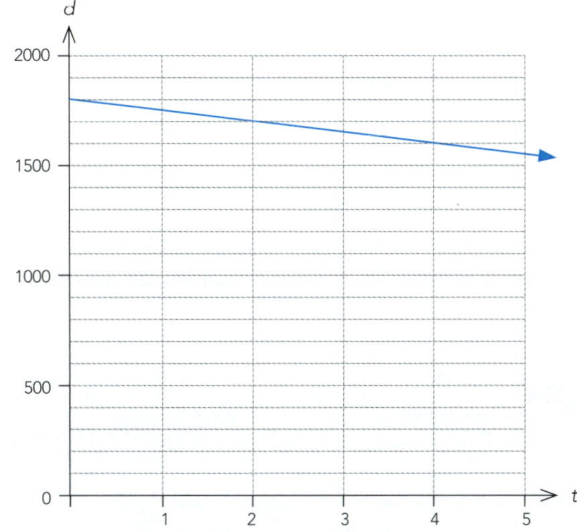

b $d = -50t + 1800$ **c** 1200 m

5 **a**

Time (min) (t)	Cost (c) (c)
2	160
4	220
6	280
8	340
10	400
12	460

b $c = 30t + 100$

c $c = 30 \times 30 + 100$

$= 1000$c

$= \$10$

d 30c per minute

ISBN: 9780170451505

e

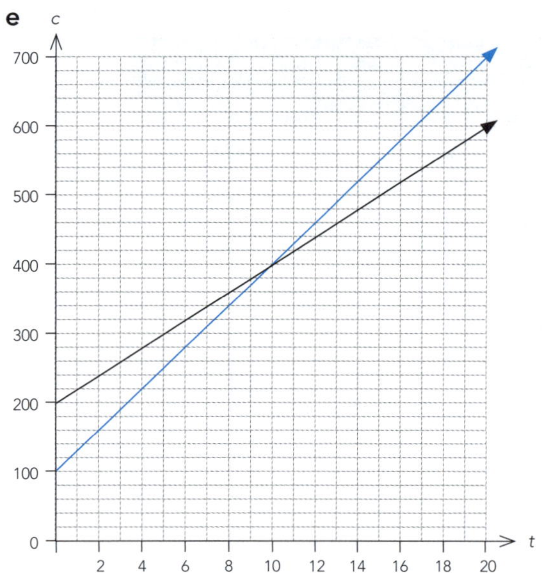

f $c = 20t + 200$

g The graph for the cost of hiring a Lemon scooter is steeper because it costs more per minute (30c) than it costs for an Orange scooter (20c).

h

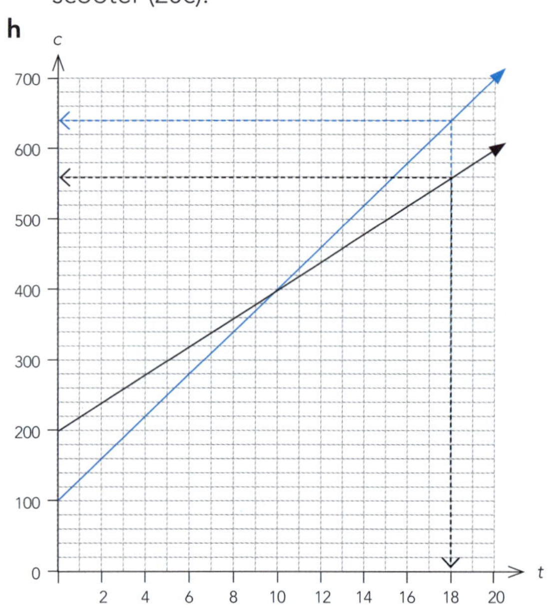

Lemon scooter: 640c = $6.40

Orange scooter: 560 c = $5.60

i It means that both scooters cost the same amount ($4.00) for 10 minutes.

Linear or not? (pp. 77–78)

1 Linear
2 Non-linear
3 Non-linear
4 Linear
5 Linear
6 Non-linear

Revision 1 (pp. 79–81)

1

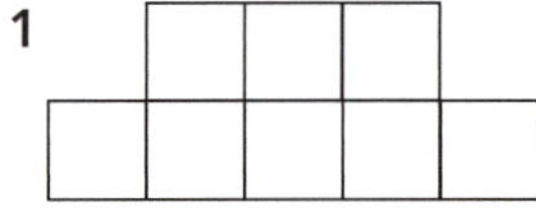

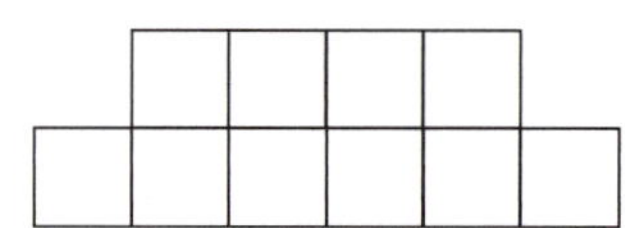

2

Shape number	Number of triangles
1	25
2	21
3	17
4	13
5	9
6	5

– 4, – 4, – 4, – 4, – 4

3 a

Term number (n)	0	1	2	3	4	5	6
Number of popsicle sticks (P)	4	9	14	19	24	29	34

b Number of popsicle sticks
= 5 x term number + 4
$P = 5n + 4$

c $P = 5 \times 50 + 4$
$= 254$

4

Term number (n)	0	1	2	3	4	5
Value of term (T)	1	3	5	7	9	11

Rule: $T = 2n + 1$

5 a

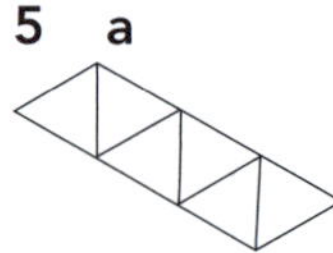

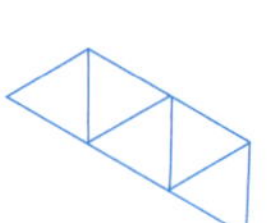

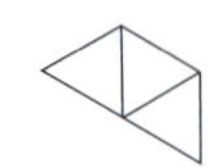

b 4, 7, 10, 13, 16, 19, …

6 a Rule $T = 5n + 2$ The 15th term = 77

b Rule $T = -3n + 24$ The 22nd term = –42

7 a

Term number (n)	1	2	3	4	5
Calculations	3 x 1 – 4				3 x 5 – 4
Value of term (V)	–1	2	5	8	11

Sequence: –1, 2, 5, 8, 11

b

Term number (x)	Calculations	Value of term (y)
1	–2 x 1 + 5	3
2		1
3		–1
4		–3
5	–2 x 5 + 5	–5

Sequence: 3, 1, –1, –3, –5

ISBN: 9780170451505

8

x	Calculation	y	Coordinates
0	2 x 0 + 1	1	(0, 1)
1		3	(1, 3)
2		5	(2, 5)
3		7	(3, 7)
4		9	(4, 9)
5		11	(5, 11)
6	2 x 6 + 1	13	(6, 13)

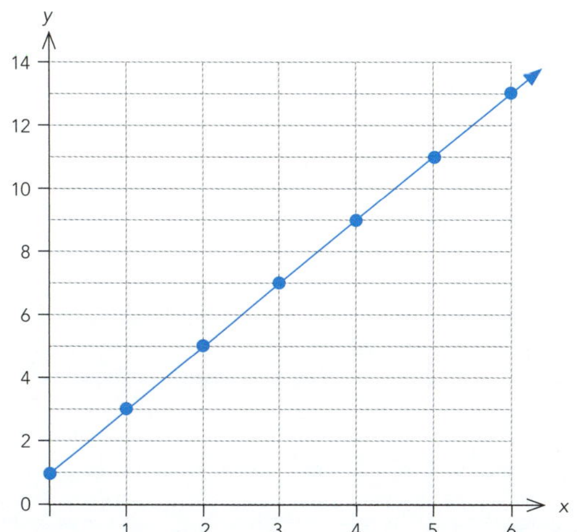

9 a

Trees (t)	Remaining netting (m) (n)
1	23
2	21
3	19
4	17
5	15

b

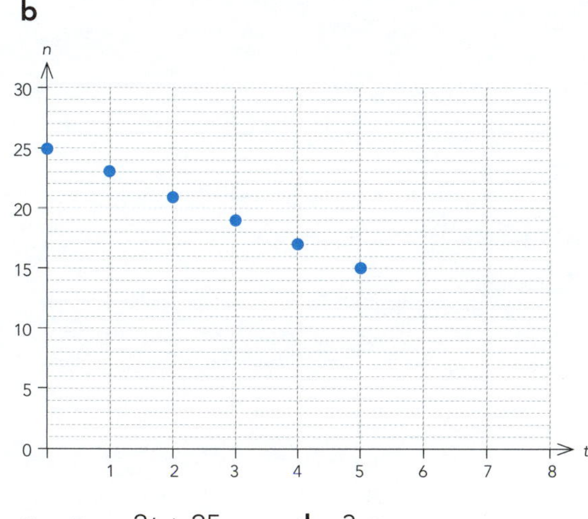

c $n = -2t + 25$ d 3 m

Revision 2 (pp. 82–84)

1

2

Shape number	Number of popsicle sticks
1	2
2	5
3	8
4	11
5	14
6	17

+ 3, + 3, + 3, + 3, + 3

3 a

Term number (n)	0	1	2	3	4	5	6
Number of stars (S)	3	5	7	9	11	13	15

b Number of stars = 2 x term number + 3
$S = 2n + 3$

c $S = 2 \times 50 + 3$
$= 103$

4

Term number (n)	0	1	2	3	4	5
Value (T)	2	5	8	11	14	17

Rule: $T = 3n + 2$

5 a

b 3, 8, 13, 18, 23, 28, 33, …

6 a Rule $T = 3n + 3$ The 21st term = 66

b Rule $T = -2n + 21$ The 15th term = –9

7 a

Term number (n)	1	2	3	4	5
Calculations	4 x 1 – 1				4 x 5 – 1
Value of term (V)	3	7	11	15	19

Sequence: 3, 7, 11, 15, 19

b

Term number (x)	Calculations	Value of term (y)
1	–3 x 1 + 6	3
2		0
3		–3
4		–6
5	–3 x 5 + 6	–9

Sequence: 3, 0, –3, –6, –9

ISBN: 9780170451505

8

x	Calculation	y	Coordinates
0	–1 x 0 + 10	10	(0, 10)
1		9	(1, 9)
2		8	(2, 8)
3		7	(3, 7)
4		6	(4, 6)
5		5	(5, 5)
6	–1 x 6 + 10	4	(6, 4)

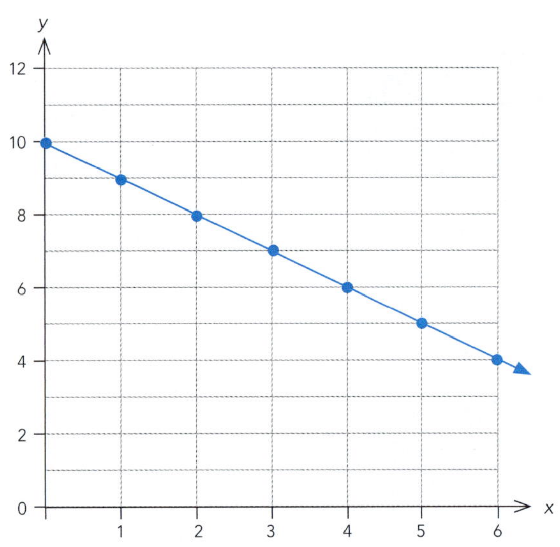

9 **a**

Cartons (c)	Piggy bank (\$) (p)
1	24
2	30
3	36
4	42
5	48

b

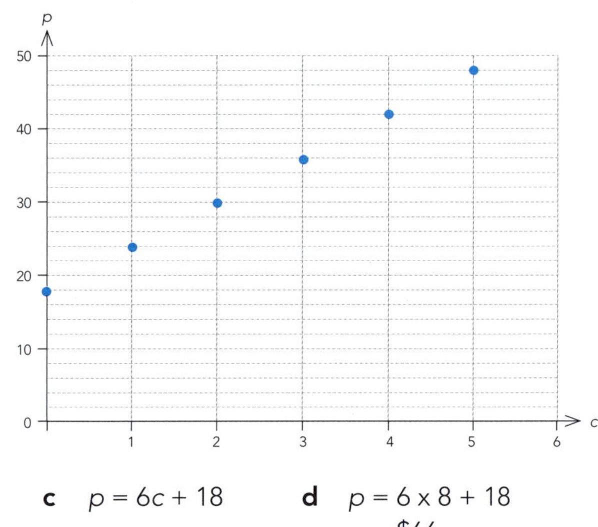

c $p = 6c + 18$

d $p = 6 \times 8 + 18$
$= \$66$

 ISBN: 9780170451505